The Politics of Climate Change

Climate change is now a mainstream part of the international political agenda. It has become clear that it is not solely a technical issue, to be resolved by scientists, but a political issue with political implications at all levels of global governance. Indeed, some may argue that few long-term problems in international affairs are more important than this one. The purpose of this book is to reveal and apply some of the latest thinking on the implications of climate change for international affairs, and to explore how various proposals for tackling climate change will affect interstate relations in coming years. Chapters by scholars of international relations, international political economy and international law contribute to current discussions of climate change, doing so in way that is accessible to students, stakeholders, government officials and informed laypersons.

Some questions considered in the book include the following: How has the discussion of climate change affected interstate relations? How does this problem, and how do environmental issues more generally, challenge international relations theory? How do international climate politics influence domestic politics, and vice-versa?

This book was previously published as a special issue of the *Cambridge Review of International Affairs*.

Paul G. Harris (PhD, Brandeis) is Chair Professor of Environmental Studies at the Hong Kong Institute of Education. From 2000 to 2009 he taught at Lingnan University, Hong Kong, where he was Professor of International and Environmental Studies, Director of the Project on Environmental Change and Foreign Policy, Director of the Environmental Studies Program, and Director of the Centre for Asian Pacific Studies. During the 1990s he was a faculty member at universities in Britain and the United States.
He is author or editor of a dozen books on global environmental policy and ethics, most recently *World Ethics and Climate Change: From International to Global Justice* (Edinburgh University Press).

The Politics of Climate Change

Environmental Dynamics in International Affairs

Edited by Paul G. Harris

Routledge
Taylor & Francis Group
LONDON AND NEW YORK

First published 2009 by Routledge
2 Park Square, Milton Park, Abingdon, Oxon, OX14 4RN

Simultaneously published in the USA and Canada
by Routledge
605 Third Avenue, New York, NY 10017

Routledge is an imprint of the Taylor & Francis Group, an informa business

Typeset in Palatino by Value Chain, India

British Library Cataloguing in Publication Data
A catalogue record for this book is available from the British Library

ISBN13: 978-0-415-48646-0 (hbk)

CONTENTS

Notes on contributors

Liliana B Andonova (PhD, Harvard University) is Associate Professor in Political Science at the Graduate Institute for International and Development Studies in Geneva, Switzerland; on leave from Colby College, USA. Previously she has been awarded fellowship positions at the European University Institute, at the Earth Institute of Columbia University and at the Belfer Center for Science and International Affairs of Harvard University. She is the author of *Transnational politics of the environment: the European Union and environmental policies in Central and Eastern Europe* (2003). Her current research examines international institutions, public-private partnerships and climate change politics.

Tom Athanasiou (MPhil, Western Institute for Social Research) is a long-time left green critic. These days, he focuses on the development and promotion of climate stabilization strategies that are fair enough to actually work. More generally, his interests focus on class division and distributive justice within finite environmental spaces. He is the author of *Divided planet: the ecology of rich and poor* (1998) and, with Paul Baer, the co-author of *Dead heat: global justice and global warming* (2002).

Paul Baer (PhD, University of California Berkley) is a Postdoctoral Scholar at the Woods Institute for the Environment at Stanford University and Research Director of EcoEquity. He received his PhD from the Energy and Resources Group at UC Berkeley in 2005, and has written extensively on topics of ethics, precaution and risk with a focus on climate change policy.

Richard Benwell (MSt, University of Cambridge) is a first year PhD candidate at the Centre of International Studies, University of Cambridge, where he is graduate coordinator of the International Political Economy Research Group. He is Parliamentary Researcher for Baroness Miller of Chilthorne Domer, with whom he has worked on the Liberal Democrat DEFRA and Home Affairs teams.

Loren R Cass (PhD, Brandeis University) is an Associate Professor of political science at the College of the Holy Cross in Worcester, Massachusetts. His research focuses on environmental foreign policy and global environmental politics. He is the author of *The failures of American and European climate policy: international norms, domestic politics, and unachievable commitments* (2006).

Oriol Costa (PhD, Autonomous University of Barcelona) is a visiting Postdoctoral Researcher at the Centre for European Integration (Otto-Suhr-Institut for Political Science), at the Freie Universität Berlin. He is also a lecturer in International Relations at the Autonomous University of Barcelona (currently on leave). Recent publications include articles in *South European Society and Politics, Revista Española* *de Derecho Internacional* and *Política Exterior*, as well as volumes edited by Esther Barbé and Paul Harris (forthcoming).

Glenn Fieldman (PhD, University of Denver) is an Assistant Professor of environmental studies at San Francisco State University and is a member of the advisory board of EcoEquity.

Paul G. Harris (PhD, Brandeis) is Chair Professor of Environmental Studies at the Hong Kong Institute of Education. From 2000 to 2009 he taught at Lingnan University, Hong Kong, where he was Professor of International and Environmental Studies, Director of the Project on Environmental Change and Foreign Policy, Director of the Environmental Studies Program, and Director of the Centre for Asian Pacific Studies. During the 1990s he was a faculty member at universities in Britain and the United States.

He is author or editor of a dozen books on global environmental policy and ethics, most recently *World Ethics and Climate Change: From International to Global Justice* (Edinburgh University Press).

Sivan Kartha (PhD, Cornell University) is a Senior Scientist at the Stockholm Environment Institute whose research and publications pertain to assessments of energy technologies and policy analyses relating to climate change and sustainable development. As Director of SEI's Climate and Energy Programme, he focuses on equity and efficiency in the formation of an international climate regime, and has worked with policy-makers, private sector actors, foundations and NGOs around the world.

Stephan Kroll (PhD, University of Wyoming) is Assistant Professor in the Department of Agricultural and Resource Economics at Colorado State University. His research focuses on the behavioural and institutional components of environmental decision-making. He has published in *Economic Inquiry, Experimental Economics, Journal of Conflict Resolution, Journal of Economic and Behavioral Organization, Public Choice, Economics Letters* and *Theory and Decision*. He recently co-edited the volume *Environmental economics, experimental methods* (2007).

Bradley C Parks (MSc, London School of Economics and Political Science) is Associate Director in the Department of Policy and International Relations at the Millennium Challenge Corporation and a research fellow at the College of William and Mary's Institute for the Theory and Practice of International Relations. He has written and contributed to several books and articles on global environmental politics, international political economy, and development theory and practice. His latest publication is *Greening aid? Understanding the environmental impact of development assistance* (2008).

J Timmons Roberts (PhD, Johns Hopkins University) is Chancellor Professor of Sociology and Interim Director of Environmental Science and Policy at the College of William and Mary, USA. He taught at Tulane University in New Orleans for ten years where he held a joint appointment in Latin American Studies and Sociology and was Co-Director of Environmental Studies. Recent co-authored books include *Greening aid? Understanding the environmental impact of development assistance* (2008) and *A climate of injustice: global inequality, North-South politics, and climate policy* (2007).

Miriam Schroeder (MA, University of Tuebingen) is a Research Fellow at Potsdam University and works for the project 'Emerging Modes of Governance and Climate Protection: Green Companies in Newly Industrializing Countries' funded by the German Research Foundation (DFG). Her research focuses on theories of International Relations and on Chinese climate politics.

Shirley V Scott (PhD, University of Queensland) is an Associate Professor of International Relations at the University of New South Wales in Sydney, Australia. She has published extensively on aspects of the relationship between international law and world politics in journals of both International Law and International Relations. She is the author of *International law in world politics: an introduction* (2004) and *The political interpretation of multilateral treaties* (2004) and editor of *International law and politics: key documents* (2006).

Jason F Shogren (PhD, University of Wyoming) is the Stroock Professor of natural resource conservation and management at the University of Wyoming and the King Carl XVI Gustaf Professor of Environmental Sciences at Umeå University. He served for the Council of Economic Advisers in the White House in 1997 during the run-up to Kyoto, and was a lead author for the IPCC Third Assessment Report. He is the author of over 200 peer-reviewed articles and 16 books.

Maria Julia Trombetta (PhD, University of Aberystwyth) is a Postdoctoral Researcher, working on energy security at the Delft University of Technology (The Netherlands). She has a degree in Economic and Social Sciences from Bocconi University (Milan) and a master's in Environmental Management from the European School of Advanced Studies (University of Pavia, Italy) She has been a Fulbright scholar working on comparative environmental politics and has lectured at Oxford Brookes University.

Introduction: the glacial politics of climate change

Paul G Harris[1]

Climate change is now a mainstream part of the international political agenda. Indeed, for anyone interested in, and concerned about, international affairs, few issues are more important. Climate change is not solely a technical issue to be resolved by scientists, but a political issue with political implications at all levels of global governance. Consequently, climate change has been the subject of three decades of diplomacy, and it is now a major concern of governments, international organizations, businesses and nongovernmental organizations, as well as increasing numbers of people around the world. But as the problem has grown in prominence, so have predictions of its adverse impacts, many of which are being felt today. Political responses have dramatically failed to keep up with the accelerating pace of climate change. Given the need to act very aggressively to limit and cope with climate change and its effects, a question arises: Why has more not been done? The purpose of this book is to help answer that question by revealing and applying some of the latest thinking about climate change in international affairs, and to explore how various proposals for tackling it will affect interstate relations in coming years. In this introduction I set the stage by (1) recounting recent assessments of how climate change affects the world and (2) summarizing the history of interstate responses to those assessments.

Global warming and climate change

Over the last two decades, scientists have radically improved their understanding of the causes and consequence of global warming—the warming of the Earth as a consequence of greenhouse gases (GHGs) building up in the atmosphere.[2] The Intergovernmental Panel on Climate Change (IPCC), created by governments in 1988 to study climate change, has concluded with '*very high* confidence that the global average net effect of human activities since 1750 has been one of warming' (IPCC 2007, 37). Carbon dioxide, the most influential GHG in aggregate, is emitted through the burning of fossil fuels (for example, coal, oil, natural gas), and when trees are felled and subsequently decay or are burned. 'Climate change' refers to

[1] I am grateful to Mui Pong Goh for expert help in pulling this collection together, Jon Symons for informed proofreading of this introduction, and Robert Keohane and many anonymous referees for comments on earlier drafts of essays to follow.

[2] Here I invoke arguments in Harris (2007b).

changes in climate and their consequences resulting from global warming, with the United Nations (UN) Framework Convention on Climate Change (FCCC) including under this rubric atmospheric changes connected directly or indirectly to human activities.[3] This human-induced global warming was, until recently, viewed as a *future* problem. But it is becoming clearer that *ongoing* climatic changes are consequences of global warming ('Climate change' 2006). The impacts of climate change on natural ecosystems and on human society and economies are potentially severe, particularly in parts of the world where geographic vulnerability and poverty make adaptation difficult or impossible. Furthermore, what is important in understanding the politics of climate change is that it is intimately connected to most economic activity and modern lifestyles, thereby tying Earth systems intimately with human systems (see Figure 1).

Recent and ongoing effects of climate change

The most authoritative reports on the causes and consequences of climate change have come from the IPCC, especially its 2007 fourth assessment report.[4] According to that assessment, since 1970 anthropogenic GHG emissions have increased globally by 70 per cent—with CO_2 in particular increasing by 80 per cent since 1995. The IPCC reports that 'atmospheric concentrations of CO_2 and CH_4 [methane] in 2005 exceed by far the natural range over the last 650,000 years' (IPCC 2007, 37). The concentration of CO_2 in the atmosphere in 2005 was 379 parts per million (ppm) compared to 280 ppm prior to the Industrial Revolution, with the annual increase being nearly 2 ppm. Importantly, although plants and the oceans absorb CO_2, global warming inhibits their ability to do so, thereby creating a feedback loop contributing to more warming and greater climate change. Perhaps seeking to counter the political influence of 'climate sceptics'—who question the reality of global warming and attribute it to all manner of causes, such as sun spots—the IPCC has declared that '[w]arming of the climate system is unequivocal, as is now evident from observations of increases in global average air and ocean temperatures, widespread melting of snow and ice and rising global average sea level' (2007, 30). What is more, in a new determination since its third assessment report in 2001, the panel found that 'discernible human influences extend beyond average temperature to other aspects of climate, including temperature extremes and wind patterns' (2007, 40). That is, the impacts of climate change are undoubtedly attributable to human activities.

Among many ongoing adverse impacts of climate change, the proportion of Earth affected by drought has increased, as has the frequency of extreme weather events, heavy precipitation, incidence of intense tropical cyclones, extreme high sea levels in a wide range of locations, and heat waves (in most regions). Meanwhile, the frequency of cool days and nights has declined. These changes are having noticeable effects on both physical and biological systems, as demonstrated by melting glaciers and sea ice; warming of lakes and rivers; the

[3] 'Climate change' in IPCC lexicon refers to changes from both natural processes and human activities, whereas the FCCC addresses only the latter.

[4] Here I summarize findings of the IPCC's multivolume fourth assessment report as recounted in IPCC (2007). The full IPCC reports are available at <http://www.ipcc.ch/>.

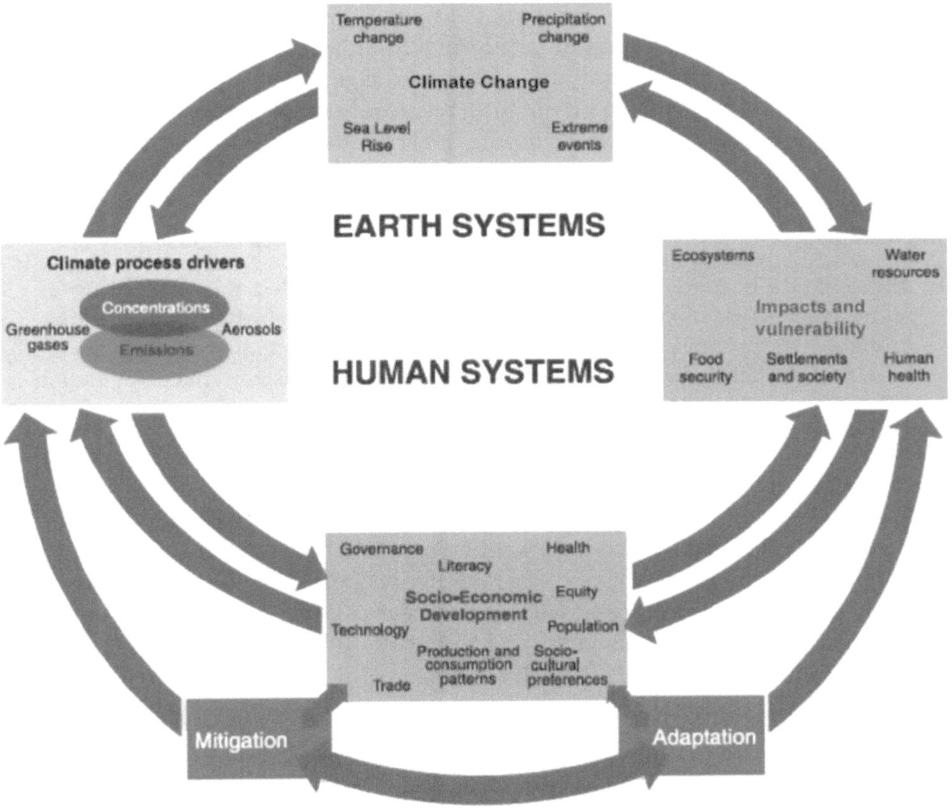

Figure 1. Anthropogenic drivers, impacts, responses and linkages (IPCC 2007, 26)
Source: Climate Change 2007: Synthesis Report. Contribution of Working Groups I, II and III to the Fourth Assessment Report of the Intergovernmental Panel on Climate Change. Figure I.1. IPCC, Geneva, Switzerland.

early advent of spring and associated changes to plants and wildlife, such as the earlier greening of vegetation and the corresponding impacts on bird migration and egg laying; and major alterations in marine ecosystems, including changes in salinity and currents, changes in the ranges of marine life and timing and locations of fish migrations, likely adverse impacts on reefs, and losses of coastal wetlands and mangroves (both crucial for healthy fisheries). The IPCC reports adverse changes to agriculture and harm to forests from more fires and pests. Human health has also been affected by heat stresses and expanding ranges of disease vectors (for instance, mosquitoes), among other effects.

Future effects of climate change

The IPCC has considered the influence of planned and likely national sustainable development policies and efforts to mitigate climate change. Its findings are not optimistic: even following the adoption of anticipated proactive policies, GHG emissions will climb. The panel projected that that there will be an increase of 0.2 degrees Celsius per decade under most emissions scenarios for the next two decades; future temperature increase will be dependent on the world's response.

Global average temperature is predicted to rise by 1.4 to 5.8 degrees Celsius, with the highest increase likely without additional mitigation policies. With continued warming, expected manifestations of climate change in this century will be 'larger' (that is, usually more adverse) than those seen in the last century (IPCC 2007, 45). Changes expected this century include generally higher temperatures over land and at high northern latitudes, reduced snow cover, thawing permafrost, shrinking sea ice, rising sea level, more frequent heat waves, heavy precipitation events and more intense tropical cyclones. As a consequence,

> the resilience of many ecosystems is likely to be exceeded this century by an unprecedented combination of climate change, associated disturbances (eg flooding, drought, wildfire, insects, ocean acidification) and other global change drivers (eg land-use change, pollution, fragmentation of natural systems, overexploitation of resources). (IPCC 2007, 48)

Positive feedbacks will increase as carbon uptake by plants reaches saturation. The risk of extinction for 20–30 per cent of plant and animal species will increase (based on only 2.5 degrees of warming), and changes in biodiversity and ecosystems seen in the last century will be exacerbated. All these will adversely affecting human needs such as water and food supplies. Coastal erosion and flooding due to sea-level rise will also increase. Extreme weather events are expected to become more frequent and intense, with 'mostly adverse effects on natural and human systems' (IPCC 2007, 53).

The health of millions of people will be adversely affected

> through, for example, increases in malnutrition; increased deaths, diseases and injury due to extreme weather events; increased burden of diarrhoeal diseases; increased frequency of cardio-respiratory diseases due to higher concentrations of ground-level ozone in urban areas related to climate change; and the altered spatial distribution of some infectious diseases. (IPCC 2007, 48)

Even in affluent parts of the world, which have a greater aggregate capacity to adapt, some groups of people, notably the poor and the elderly, will suffer the risks of climate change. The upshot is that, around the world, '[m]ore people are projected to be harmed than benefited by climate change', even if temperature increases are somehow mitigated (Working Group II 2001, paragraph 2.8).

Regional affects will vary, ranging from up to hundreds of millions of people exposed to water stress in Africa, increased flooding in the coastal and delta regions of Asia, significant loss of biodiversity in Australia, retreat of glaciers in the mountains of Europe and water shortages in southern Europe, loss of tropical forests and biodiversity in Latin America, water shortages and heat waves in North America, detrimental changes to natural ecosystems in polar regions, and inundations and storm surges in small islands—to list only a few of the anticipated changes in coming decades. Later in the century, the likelihood of abrupt or irreversible environmental changes will increase, with some of them considered inevitable. These could include rapid sea-level rise, significant extinctions (40–70 per cent of species if temperature increases exceed 3.5 degrees Celsius), large-scale, persistent changes to marine systems and fisheries, and yet more positive (in other words, harmful) feedback loops as oceans absorb more CO_2. In future centuries, impacts of climate change could be truly monumental.

The international response to climate change

Scientific assessments from the IPCC and other scientists provided the stimulus for international agreements to address climate change.[5] However, because the science has been intimately wrapped up with politics, climate diplomacy has often taken on a life of its own—one that is partly divorced from science. One of the earliest significant international events was the 1979 First World Climate Conference, a gathering of scientists interested in climate change and its relationship with human activities. From that conference a program of scientific research was established, leading to the creation of the IPCC in 1988. The IPCC's first assessment report and the Second World Climate Conference in 1990 added stimulus to the initial concerns about climate change among governments. Therefore, in December 1990, the UN General Assembly established the Intergovernmental Negotiating Committee for a Framework Convention on Climate Change. The goal of the committee was to negotiate a framework convention that would be the basis for subsequent international protocols dealing with climate change.

The Framework Convention on Climate Change and the Kyoto Protocol

From then till the 1992 UN Conference on Environment and Development ('Earth Summit'), representatives of over 150 countries negotiated the FCCC. The stated aim of the FCCC is the

> stabilization of greenhouse gas concentrations in the atmosphere at a level that would prevent dangerous anthropogenic interference with the climate system. Such a level should be achieved within a time-frame sufficient to allow ecosystems to adapt naturally to climate change, to ensure that food production is not threatened and to enable economic development to proceed in a sustainable manner. (article 2)[6]

The FCCC called on the world's most economically developed countries to reduce their emissions of GHGs to 1990 levels by 2000. However, this objective was not achieved. The FCCC came into force in 1994, after ratification by 50 countries. Particular responsibility was also laid on the developed states to provide 'new and additional' resources to developing countries to help them with their efforts to limit GHG emissions. If the negotiation of the FCCC was fraught with tensions between developed and developing countries, the negotiations after 1992 were even more contentious.

In 1995 the parties to the FCCC established the Conference of the Parties (COP), which became the convention's overriding authority. Many COP meetings were held to negotiate the details of how GHG emissions limitations would be achieved. At COP1 meeting (held in Berlin in 1995), developed countries acknowledged that they had a greater share of the responsibility for causing climate change and they would act to address it first. Central to the resulting Berlin Mandate was the demand by developing countries that the industrialized countries take on greater commitments to reduce their GHG emissions and assist

---

[5] For similar depictions of climate diplomacy, see Harris (2007a) and Harris (2007b).

[6] United Nations Framework Convention on Climate Change, 1992, reprinted in 31 ILM 84 (1992).

the poor countries with sustainable development. Thus COP1 affirmed the notion of 'common but differentiated responsibilities', meaning that, while all states have a common responsibility to address climate change, the developed countries have greater ('differentiated') obligation to do so. At COP2 (held in Geneva in 1996), countries called for a legally binding protocol with specific targets and timetables for reductions of GHG emissions by developed countries. The resulting Geneva Declaration served as the negotiating basis for the Kyoto Protocol, which was agreed in December 1997 at COP3 in Kyoto. The Protocol requires most developed country parties to reduce their aggregate GHG emissions by 5.2 per cent below 1990 levels between 2008 and 2012. However, not all developed countries agreed to be bound by the Protocol.

The Kyoto conference proved to be especially contentious, not least because the United States (US) seemed to have reneged on the Berlin Mandate when President Bill Clinton called for the 'meaningful participation' of developing countries. Nevertheless, diplomats at the conference managed to agree to the Kyoto Protocol, which established specific emissions goals for developed countries without requiring significant commitments from developing countries. The Protocol also endorsed emissions-trading programmes that would allow developed countries to buy and sell emissions credits among themselves. Other so-called flexible mechanisms included in the Protocol were 'Joint Implementation' (JI), whereby developed countries could earn emissions credits when investing in one another's emissions-reduction projects, and the Clean Development Mechanism (CDM), which allows developed country entities to pay for—and receive emissions credits for—emissions-reduction projects in developing countries.

Implementing the Kyoto Protocol

Some of the means by which the Kyoto Protocol's 5 per cent goal would be reached were codified at COP4 (held in Buenos Aires in 1998,). At COP5 in Bonn held in October 1999, the parties agreed to a timetable for completing outstanding details of the Kyoto Protocol by COP6 and, in an effort to speed up negotiations, gave the conference president the power to 'take all necessary steps to intensify the negotiating process on all issues during the coming year' (FCCC 1999). The sixth COP began in November 2000 in The Hague, but the talks broke down due to disagreements among delegates, particularly on the question of carbon sinks, which are processes, such as planting trees (afforestation) that can remove GHGs from the atmosphere. The Kyoto Protocol's ratification was put into doubt with the advent of President George W Bush in the United States, who withdrew all US support for it. The sixth COP resumed in Bonn during July 2001. The resulting Bonn Agreement clarified plans for emissions trading, carbon sinks, compliance mechanisms and aid to developing countries. At COP7 (held in Marrakech in 2001), the parties to the FCC agreed to a long list of ways to meet the Kyoto commitments. The result was the Marrakech Accords, a complicated mix of proposals for implementing the Kyoto Protocol, largely designed to garner ratification from enough states to allow the Protocol to enter into force. The parties agreed to increase funding for the FCCC's financial mechanism, the Global Environmental Facility, as well as to establish three new funds that would provide additional aid to poor countries: the Least Developed Countries Fund, the Special Climate Change Fund and the Adaptation Fund.

A milestone of sorts was reached at the October 2002 COP8 in New Delhi. A tacit agreement—between the US, a few other developed countries and several large developing countries, notably China and India—emerged, shifting much of the focus away from mitigating climate change and toward adaptation—wealthy countries agreeing to help developing countries adapt to climate change, rather than the former having to reduce their GHG pollution. It was at this COP, as well as at COP9, that diplomats discussed ways to implement the Marrakech Accords and prepare for ratification of the Kyoto Protocol. The tenth COP (held in Buenos Aires in 2004), was dubbed the 'Adaptation COP' because the discussion focused more on adaptation to climate change than the more usual COP discussions about efforts to mitigate it through emissions limitations. In the end, there were pledges for more assistance to aid poor countries most affected by climate change, but there were no firm commitments to make access to adaptation funds easier for developing countries. Importantly, it was also in 2004 that Russia ratified the Kyoto Protocol, allowing the agreement to finally enter into force in February 2005.

One visible aspect of the climate change negotiations has been the acrimony between the developed countries—particularly the US—and the developing world. The international negotiations have been plagued by efforts of developed countries to persuade developing countries to commit to emissions limitations, on the one hand, and developing country efforts to avoid such commitments on the other. These differences were manifested during the late-2005 combined COP11 and 'First Conference of the Parties Serving as the Meeting of the Parties' to the Kyoto Protocol (COP/MOP-1), which were held simultaneously in Montreal. Despite US attempts to derail the meeting, it formalized rules for implementing the Protocol (for example, rules for emissions trading, joint implementation, crediting of emissions sinks and penalties for noncompliance), streamlined and strengthened the CDM, began negotiations for further commitments by developed country parties to the Protocol beyond 2012 (when the Kyoto commitments expire), set out guidelines for an Adaptation Fund, and initiated a process for negotiating long-term action to combat climate change. Several developing countries, although still opposed to binding obligations, showed new interest in undertaking voluntary measures, in keeping with the principle of common but differentiated responsibility.

Building on the Kyoto Protocol?

Climate negotiations in the last few years have resulted in mostly incremental progress. In his opening address to the November 2006 COP12 in Nairobi, UN Secretary-General Kofi Annan characterized the negotiations up to that point as displaying a 'frightening lack of leadership' from governments (Annan 2006). At COP13 (held in Bali in 2007), the familiar arguments between developed and developing countries were manifest: European states argued in favour of deeper international commitments for GHG cuts, the US strongly opposed them, and developing countries argued for more financial and technological assistance (Pew Centre 2007). The discussions at Bali were pushed to a substantial degree by the IPCC's fourth assessment report, which removed any remaining doubt (among officials willing to entertain the facts) about the seriousness of the problem. The meeting was important in its widespread opposition to efforts by US diplomats to thwart negotiations on a new, post-2012 agreement that would obligate developed countries to take on new responsibilities to limit GHG emissions and aid developing countries with sustainable development. In the end, developing country governments agreed that they would consider taking unspecified future actions to mitigate their GHG emissions, which was a substantial shift from their longstanding policy of refusing to agree to any commitments whatsoever. The *quid pro quo* for the developing countries' stated willingness to consider future emissions limitations was the streamlining of the Adaptation Fund and financing it with a new 2 per cent levy on CDM projects. Developed countries also agreed to new emissions targets and timetables—but, as with the developing states' agreement, nothing was specified. Diplomats instead adopted the so-called Bali Roadmap, intended to guide discussions leading to a new, comprehensive agreement, under both the FCCC and the Kyoto Protocol, to be agreed in time for a conference of parties in Copenhagen at the end of 2009.

All of these international efforts to address climate change have been far too timid when viewed in relation to the severity of the problem. Even with full implementation, the Kyoto Protocol will result in reductions of well under 5 per cent of the parties' emissions because the manner in which they are allowed to meet their commitments (for example, emissions trading and land-use changes) often will not result in significant national emissions cuts. However, scientists tell us that the emissions of CO_2 must be ended *completely* just to stabilize their concentrations in the atmosphere and to prevent chaos in the global climate system (Mathews and Caldeira 2008). In what could become a seminal essay on climate change, James Hansen et al (2008) have shown that, due to the time lag before the full impact of emissions is felt, even the *current* concentration of CO_2 in the atmosphere is likely to bring dangerous interference with Earth's climate system that the FCCC was intended to prevent. Even the relatively ambitious aim of the European Union to keep global temperatures to only 2 degrees Celsius above the preindustrial levels is far too weak a target. The current concentration of CO_2—about 385 ppm—is 'already too high to maintain the climate to which humanity, wildlife, and the rest of the biosphere are adapted' (Hansen et al 2008, 15). Instead, what is required at minimum is an effort to bring CO_2 concentrations down, very quickly, to about 350 ppm, meaning a near-total move away from any use of fossil fuels if carbon cannot be captured and permanently stored—something that is not practically possible at present. According to Hansen et al,

> Present policies, with continued construction of coal-fired power plants without CO_2 capture, suggest that decision-makers do not appreciate the gravity of the situation. We must begin to move now toward the era beyond fossil fuels. Continued growth of greenhouse gas emissions, for just another decade, practically eliminate the possibility of near-term return of atmospheric composition beneath the tipping level for catastrophic effects. (15)

Consequently, the Kyoto Protocol is, at best, a very tiny step towards greater action. In the meantime, global GHG emissions will continue to rise precipitously, notably because large developing countries (especially China and India) will be increasing their use of fossil fuels as their economies grow. Climate change will continue, virtually unabated, short of new, *much more aggressive* collective action to reduce GHGs. However, strong signals of the more robust action needed are distinct in their absence. The IPCC, in a typically guarded understatement, characterizes the failure of the Kyoto Protocol this way: '[t]o be more environmentally effective, future mitigation efforts would need to achieve deeper reductions [than the Protocol] covering a higher share of global emissions' (IPCC 2007, 62). Indeed, the international legal instruments intended to avert dangerous interference with the Earth's climate—the stated aim of the FCCC—are increasingly focusing on mitigating and adapting to that dangerous interference, rather than averting it.

Conclusion

The international political response to climate change has been truly glacial. In the words of Hansen et al (2008, 15), 'the stakes, for all life on the planet, surpass those of any previous crisis. The greatest danger is continued ignorance and denial, which could make tragic consequences unavoidable.' Although some observers believe that climate diplomacy and the agreements that have resulted and that are now being negotiated are remarkable, and important steps towards the kind of action that scientists say is required have been taken, it is perhaps more accurate to describe those steps—relative to the scale of the problem and its anticipated impacts on people, communities and other life on Earth—as glacial. There has been movement in the right direction, but it is shamefully small given what is required. Even as ideas about the threat from climate change (for instance, 'climate security') and its ethical implications (for example 'climate justice') have taken hold and started to shape politics domestically and internationally, the response has been far too modest. As such, climate change demonstrates many of the limits of international politics.

Why this glacial response? Why have governments, international organizations and other important international actors started to move on climate change, and—most importantly—why have they not done much more? What can we learn from action so far to point us in new directions and new solutions? It is to these questions that the following chapters turn, and to which I return in the Conclusion.

References

Annan, Kofi (2006) 'Citing "frightening lack of leadership" on climate change', <http://www.un.org/News/Press/docs/2006/sgsm10739.doc.htm>, accessed 21 July 2008

'Climate change is all around us' (2006) *New Scientist*, 2543, March, <http://environment.newscientist.com/channel/earth/mg18925432.600-editorial-climate-change-is-all-around-us.html>, accessed 18 July 2008

Framework Convention on Climate Change (FCCC) (1999) 'Ministers pledge to finalize Climate Agreement by November 2000', Press Release, 5 November, <http://www.unis.unvienna.org/unis/pressrels/1999/env75.html>, accessed 2 October 2008

Hansen, James, et al (2008) 'Target atmospheric CO_2: where should humanity aim?', *Open Atmospheric Science Journal*, 804, April, <http://arxiv.org/pdf/0804.1126v2>, accessed 18 July 2008

Harris, Paul G (2007a) 'Collective action on climate change: the logic of regime failure', *Natural Resources Journal*, 47:1, 195–224

Harris, Paul G (ed) (2007b) *Europe and global climate change: politics, foreign policy, and regional cooperation* (Cheltenham, United Kingdom: Edward Elgar Publishing)

IPCC (2007) *Climate change 2007: synthesis report*, Contribution of Working Groups I, II and III to the Fourth Assessment Report of the Intergovernmental Panel on Climate Change <http://www.ipcc.ch/ipccreports/ar4-syr.htm>, accessed 18 July 2008

Mathews, H Damon and Ken Caldeira (2008) 'Stabilizing climate requires near-zero emissions', *Geophysical Research Letters*, 35:4, <http://www.agu.org/pubs/crossref/2008/2007GL032388.shtml>, accessed 22 July 2008

Pew Center on Global Climate Change (2007) *Summary of COP13*, <http://www.pewclimate.org/docUploads/Pew%20Center_COP%2013%20Summary.pdf>, accessed 21 July 2008

Working Group II (2001) *Climate change 2001: working group II: impacts, adaptation and vulnerability* (Cambridge: Cambridge University Press), <http://www.grida.no/CLIMATE/IPCC_TAR/wg2/010.htm>, accessed 18 July 2008

A climate of obstinacy: symbolic politics in Australian and Canadian policy

Loren R Cass

Introduction

Despite sharing a large number of attributes, Canada and Australia's foreign policy positions in international climate negotiations have varied significantly. Both are former British colonies with parliamentary governments, continent-sized countries with significant portions of their economies devoted to natural resource extraction and processing, and substantial emitters of greenhouse gas (GHG) emissions. Up until December 2007, the United States (US) and Australia were alone among the developed states in refusing to be bound by the Kyoto Protocol. Canada has ratified the Protocol but its emissions have continued to rise precipitously and it will be unable to meet its Kyoto commitments without generous use of the flexibility mechanisms contained in the Protocol. Both countries have struggled to reduce their GHG emissions, but Canada continued to pledge its support for Kyoto, whereas Australia rejected it until 3 December 2007 when Kevin Rudd—in his first official action after being sworn in as Australia's new prime minister—signed the instrument of ratification to make the Kyoto Protocol binding for Australia.

How do we explain the variation in Canadian and Australian foreign policy responses to climate change? A rational choice approach focusing on relative costs and benefits of participation in Kyoto could potentially offer an explanation. If this were the case, there should be significant variation in the costs associated with reducing GHG emissions and/or the anticipated costs or benefits of climate change. Another explanation would be to focus on the domestic political situation in each country and the relative political salience of climate change. The positions adopted by each country could merely reflect domestic political imperatives. If this were the case, there should be variation in public opinion polls. Significant changes in foreign and domestic policies should cluster around election cycles and changes in party control of government. Although both these variables have substantial explanatory value, the national responses to climate change in Australia and Canada appear to be significantly shaped by ideational forces. In particular, political leaders have pursued symbolic international climate policies to cultivate national identities for two distinct audiences—international partners and domestic voters. International climate policy positions were frequently used as signalling tools by both states, with the resulting climate positions bearing little relationship to each government's domestic climate policies.

This article is divided into three primary sections. The first section briefly discusses processes of identity formation and the role of symbolic politics in managing state identities. The second section analyses the reasons that environmental policy, and climate policy in particular, has been conducive to symbolic politics. The final section provides brief case studies of Canadian and Australian climate change policies in the light of the discussion of symbolic politics.

State identity and interests

This analysis falls within the constructivist international relations literature on state identity and international norms, defined as collective expectations about the appropriate behaviour for a given actor. It is important to note that material forces and the domestic political context heavily influenced the evolution of Canadian and Australian climate policy. However the management of state identities influenced the definition of national interests and their strategic pursuit in ways that cannot be explained purely through an analysis of material forces and domestic politics.

There is a two-level component to the management of national identities and the promotion/acceptance of associated norms of behaviour (Katzenstein 1996; Finnemore and Sikkink 1998). Identities are formed as an outgrowth of national norms and values, and they are also formed in relationship to other international actors and international normative structures. Political leaders seek to internationalize domestic norms to limit domestic adjustment costs and to extend domestic values to the international system. They also face pressure to internalize norms that have gained currency internationally. To achieve international acceptance, emergent norms must be framed in a manner seen as legitimate and appropriate by other actors within the international system. As Payne notes, 'frames provide a singular interpretation of a particular situation and then indicate appropriate behaviour for the context' (2001, 39).

Increasingly, states have sought to establish themselves as 'normative powers', defined as international actors capable of framing international opinion in a given issue area and convincing others of the rectitude of their understanding of the problem and proposed responses. It is the ability of an actor to shape conceptions of what is 'normal' in international affairs (Manners 2002, 239). Normative powers are able to influence the values of other actors—not through military force or economic pressures—but rather through the appeal of the norms themselves (Diez 2005).

When evaluating the role of Canada and Australia in international climate negotiations, important questions arise regarding the role of 'middle powers' in international relations. 'Middle powers' have been defined in a variety of ways ranging from relative economic power to physical location (Holbraad 1984; Wood 1988; Cooper et al 1993). Regardless of the definitions used, Canada and Australia are typically identified as middle powers with influence in international relations, though without the power to impose their will on reluctant partners. The literature on middle powers often characterizes these states as being relatively trustworthy because they tend to behave differently than the major powers (Holbraad 1984). For example, they tend to emphasize multilateral solutions, pursue compromises to conflicts and embrace 'good international citizenship' (Wood 1988; Cooper et al 1993). A reputation for good international citizenship is critical for middle powers because it is what creates credibility in the minds of other actors and provides them with influence in negotiations. Middle powers typically seek to promote norms related to multilateralism, wide consultation, consensus and respect for international law because their influence depends upon a rule-based system of bargaining to secure foreign policy autonomy and to create an environment more conducive to achieving their interests. The much less desirable alternative for middle powers is to depend upon larger states to protect their interests, which will inevitably place limitations on the middle power's autonomy in foreign policy.

Canada's foreign policy has been shaped by its middle power status, but it is also significantly influenced by its position relative to the US. Canada has long sought to differentiate itself from its southern neighbour to avoid being subsumed into the American foreign policy agenda. As Cooper et al note, 'Proximity to the United States should never be underestimated in helping to explain the strongly internationalist flavour to much Canadian policy' (1993, 169). Canada is dependent upon the US economically and its security is closely related to that of the US, so it cannot afford to alienate its neighbour on first tier security and economic issues. However, second tier diplomatic issues provide an important opportunity to distinguish Canadian values and related foreign policy positions from those of the US. Aligning with European positions on issues related to human rights and environmental policy provides opportunities to differentiate Canadian policy from US policy and to cultivate relations with the European Union as a counterweight to American influence.

As a middle power, Australia also finds itself in a similar structural position; its foreign policy has been shaped significantly by its position relative to its Asian neighbours and the major powers of the world. Australia also struggles with two competing visions of its international identity (Harries 2003). One vision portrays it as an 'isolated outpost and bastion of Western civilization' (Jones and Benvenuti 2006, 111). From this perspective, Australia is a beacon of stability and western values for Asia, but is also dependent upon more powerful western allies (particularly the US) to maintain regional security. The perceived dependence on the US for security and the post-Second World War reliance on the US for commerce and foreign direct investment created difficulties for Australia in defining its identity. Many Australian scholars in the 1970s and 1980s argued that the US was merely the next imperialist power to exert its will and influence over their country (Rolfe 1997). Australia has thus struggled with some of the same concerns as Canada in defining its identity *vis-à-vis* the US.

The other vision portrays Australia as a mature independent nation capable of engaging the Asian countries on their own terms. This vision sees an Australia that is fully engaged with the Asian realm; an Australia characterized by economic growth, political stability and close relations with its Asian neighbours (Jones and Benvenuti 2006). This vision found its fullest expression in the 1990s under Paul Keating's Labor Party government, which emphasized engagement with Asian neighbours and a downplaying of relations with outside powers (Kelly 2003). John Howard's Liberal/National Party coalition governments (1996–2007) were even more forceful in emphasizing sovereignty to the point of largely rejecting multilateral institutional interference in Australian affairs. They also emphasized the importance of the American-Australian relationship (Reus-Smit 2002, 6). Under Howard, the emphasis on Australia's identity and interests has been on the shared values between the US and Australia (Devetak and True 2006, 249–250). Howard has argued that '[t]he relationship we have with the United States is the most important we have with any single country. This is not only because of the strategic economic and diplomatic power of the United States. But of equal, if not more significance, are the values and aspirations we share' (Howard 2001). Howard's governments have been much more forceful in aligning Australian foreign policy with American policy. There is a tension among the major political parties regarding Australia's place in the international system and how to best pursue its interests within that system.

Both Canada and Australia partially define their identities in reference to the US, but Canada is more likely to pursue a multilateralist foreign policy agenda and emphasize its differences with the US on second tier issues. Australia, on the other hand, is more likely to support its image of good citizenship while minimizing its differences with the US—at times even emphasizing the alignment of its interests with American interests.

Environmental policy as realm of symbolic politics

In seeking to cultivate a particular national identity, states typically appeal to norms consistent with that identity. Symbolic politics can play an important role in demonstrating state support for international norms and for managing a state's identity in the eyes of both its international partners and its domestic public. The concept of symbolic politics can be traced to Edelman (1964). Symbolic politics can be defined as:

> a political process, in which certain goals and measures are announced and enforced, which already at the very early stage of publication either represent sheer rhetoric and thus only target a signalling effect or are designed in such a way that these goals and measures should or could not be realized and implemented in the same way as they are announced. (Hansjurgens 2000, 147, cited in Matten 2003, 216)

Blühdorn (2007) argues that there are two primary forms of symbolic politics that can be distinguished depending on the intention of the actors utilizing symbolic politics. The first form of symbolic politics relates to the political use of symbols to communicate with other policy-makers and the broader public. This form of symbolic politics helps actors to mobilize segments of society, to reduce the complexity of an issue and to generate broader support for the policies advocated by the actor (Blühdorn 2007): '[S]ymbolic politics in this understanding always entails the strategic construction of realities and manipulation of mass audiences' (256). At the international level, this strategy may be used to help establish an international identity with the intent of altering domestic policy and/or influencing international negotiations to convince the target audience of the need to change. In this sense, symbolic politics is a communication strategy that is indispensable to shaping the public and/or international debate in a given issue area.

The alternative form of symbolic politics is more insidious. In this form, symbolic politics represents the substitution of symbolic acts for real action. In other words, symbolic actions are undertaken not in an attempt to communicate more effectively with the target audience but rather to deceive that audience and lead it to believe that actions are being undertaken to address a concern when in fact the symbolic actions will not have the stated or implied effect in practice. Both forms of symbolic politics are present in these case studies, but the second form is much more prominent.

When will political actors pursue symbolic politics? Newig (2007) hypothesizes that four conditions tend to promote the use of symbolic measures. Firstly, when there is high political pressure, decision-makers feel compelled to do something to address the problem. This creates a political incentive to utilize symbolic politics. Secondly, if the short-term costs of addressing the problem are high relative to the short-term benefits or if there are no viable policy options available, decision-makers are likely to resort to symbolic politics. Thirdly, higher levels of conflict among groups within society will tend to promote symbolic legislation. Finally, greater complexity in an issue and the asymmetrical availability of information tends to promote symbolic legislation (282).

Environmental policy is a particularly ripe field for the pursuit of symbolic politics. Many environmental problems generate public demands for action, but policies to address them frequently involve substantial short-term costs with often only long-term benefits. Costs are often concentrated on a small number of actors, whereas the benefits are more diffuse. Most environmental issues are extremely complex, involving substantial scientific and policy-related uncertainties. In addition, environmental policy shares a number of attributes with the field of human rights that generate demands for action on moral grounds. As Sikkink notes, 'Human rights norms have a special status because they prescribe rules for appropriate behaviour and they help define identities of liberal states' (1998, 520). The 1972 Stockholm Conference, the 1972 publication of the Club of Rome's *Limits to Growth*, the 1987 publication of *Our Common Future*, the 1992 and 2002 Earth Summits, among numerous other events and publications, have produced a growing focus on 'sustainable development' as the dominant paradigm for the pursuit of economic development consistent with needs of future generations. Increasingly, support for sustainable development has become one of the touchstones of maintaining a state's identity as a legitimate member of the society of liberal states. Although debates abound regarding what sustainable development means, in practice reference to the term has become essential to providing legitimacy to a proposed environmental policy (Bernstein 2002).

The emphasis on sustainable development and the needs of developing countries make it easier to frame global environmental problems as harming vulnerable populations around the world. To be a legitimate member of the international system thus requires acceptance of the need to act to address these problems. The need to support action does not necessarily mean that states will fulfil the domestic or foreign policy changes implied by those commitments. Environmental policies permit states to take positions consistent with international good citizenship with the knowledge that international enforcement remains extremely weak. As a characteristically secondary issue area in international relations, environmental policy is also an area in which states can pursue policies in opposition to dominant international actors without fear of significant retribution. The need for cooperation among a large number of states also creates niche opportunities for small and medium-sized states to exert significant influence on shaping international negotiations (Cooper 1997; Kanie 2007). We would thus expect smaller states to be more active in issue areas like human rights and environmental politics where these states can pursue policies to establish their identities as good international citizens without significant political costs.

Middle powers have a strong incentive to align their foreign policies with liberal international norms. Lacking the power to forcefully achieve their interests, they depend upon the maintenance of a rule-based international system and their reputations as legitimate, influential members of that system to achieve their interests. The reputations of middle powers are linked to the perception of these states as trustworthy and willing to comply with their international obligations. A middle state will face a particularly strong incentive to affirm an emergent international norm if there is substantial agreement on its appropriateness among states with which the middle state identifies.

To the extent that international commitments are consistent with domestic political incentive structures, the government can pursue policies at both levels to cultivate its identity as a legitimate actor in the international system. However, there is a potential tension between international norms and domestic support for the requisite behavioural changes. Cases in which international requirements for action come into conflict with domestic political priorities are ripe for the pursuit of symbolic politics. Symbolic foreign policy positions in support of international norms but without pursuit of the implied domestic policy changes permit the government to cultivate an identity as a legitimate supporter of the international system without facing substantial adverse domestic policy consequences. The government can also communicate its support for the broader group of states promoting the norm, which may have additional benefits in other issue areas. To the extent that there is domestic support for the policies, such a position could also enhance the government's standing in these constituencies without alienating actors who could be harmed by domestic policy changes. The pursuit of symbolic foreign policy positions without the requisite domestic policy changes is sustainable only as long as there is insufficient information and motivation for the public and/or international actors to evaluate the government's performance and hold it accountable. Presumably, the government will eventually be forced to reconcile its domestic and international policies or face some form of retribution.

The alternative situation is one in which the government opposes the emergent norms and/or associated policies. Again, middle powers face substantial pressure to affirm the emergent norm to maintain their identities as legitimate members of the international system. However, if there is sufficient international opposition to the emergent norm, a middle power may seek to align itself with the opposing group in an attempt to undermine the norm and/or to symbolically communicate its support for the other states. Such a policy becomes more difficult if there is substantial domestic support for the international norm. In this case, the government may pursue domestic policies consistent with the international norm even as it rejects them in its foreign policy to align itself with other countries internationally.

The management of national identities in the international system was particularly important in the late 1980s and early 1990s. With the end of the Cold War and the relative decline in direct security threats to the global order, middle powers found new opportunities to seek out leadership positions in international negotiations (Cooper et al 1993). Climate change provided opportunities for Canada and Australia to pursue symbolic politics to help refine post-Cold War identities and to signal priorities to both domestic and international audiences. The high profile nature of climate change after the mid-1980s made it particularly attractive for symbolic politics. A change in policy could immediately communicate a shift in priorities or values to actors at home and abroad. Both countries pursued international climate policy positions that were largely symbolic in nature. The international positions typically did not result in domestic policy changes; instead they were designed to manage national identities and to signal shifts in government priorities. The governments of both countries recognized that the early commitments to action would not entail immediate demands for domestic policy changes. However, over time it would become increasingly difficult to separate international and domestic policies. The tension between the two levels and the use of symbolic politics to manage that tension explains some of the variation in national responses to climate change.

The politics of climate change in Canada and Australia

The political evolution of the international response to climate change covers the period from the late 1970s up through the present. There is insufficient space here to address the political responses of Canada and Australia in detail over the entire period. Instead, the focus will be on transition points in each country's foreign policy responses to climate change. Each country staked out an initial position in the late 1980s and then had to re-evaluate those positions in response to the evolving international negotiations. The 1990–1992 United Nations Framework Convention on Climate Change (UNFCCC) negotiations, the 1995–1997 Kyoto Protocol negotiations over specific national commitments/policy changes and the 2001 American rejection of Kyoto created additional foreign policy decision points for Canada and Australia, and their responses varied. How do we explain the evolution of Canadian and Australian foreign policy during these periods?

Material forces in Australian and Canadian climate policy

The potentially most straightforward explanation for Australian and Canadian climate policy could be differences in economic incentives. In 1990, the UNFCCC/Kyoto Protocol base year for calculations of GHG emissions, the US was the largest emitter of carbon dioxide. Canada and Australia were the eighth and eleventh largest emitters among the developed states, and their emission patterns were similar. Their per capita emissions were nearly twice the Organisation for Economic Co-operation and Development (OECD) average and more than twice those of the major European economies. Their carbon emissions per dollar of Gross Domestic Product (GDP) were also substantially higher than the OECD and European rates. In part, the higher emission rates reflect the greater dependence of these economies on resource extraction and processing as well as their continental sizes and low population densities, which result in substantially higher emissions for transportation.

Böhringer and Vogt (2004) use a political economy analysis to demonstrate why the US opted out of the Kyoto Protocol. They argue that the relatively high compliance costs and the low willingness of the American public to pay for emission reductions made it rational for political leaders interested in re-election to walk away from the Kyoto Protocol. However, evaluating the cases of Canada and Australia in the same model produces some anomalies. The cost of abatement in both countries was relatively high. The median marginal abatement costs per ton of carbon dioxide were higher in Canada at US$145 compared to Australia's US$124. Both countries were substantially higher than Europe's average of US$110 per ton (610). This implies that Canadians would have to be willing to pay even higher costs of abatement than the Australians. However, in 1998 and 2000 surveys, 53.7 per cent and 49.0 per cent of respondents in Australia saw climate change as 'very serious'. In Canada the results were 43.5 per cent and 44.6 per cent (607). This suggests that Australians' willingness to pay was likely to be similar if not higher than Canadians'. However Australia opted out of the Kyoto Protocol and Canada signed it. Once the US opted out of Kyoto and full emissions trading was permitted, the marginal costs of abatement were potentially the same for all countries because states could purchase emission reduction credits abroad if domestic emission reductions proved too costly. The marginal costs of abatement would have then fallen to US$42 per tonne based on Böhringer and Vogt's model (2004, 610). As such, the material incentives facing both countries should have been similar. Thus, it would appear that a pure political economy approach cannot fully explain the policy choices. The explanation must lie in domestic political differences and/or in the ideational variables outlined in the prior sections.

Canadian climate policy

Although a material-based approach cannot fully explain Canada's policy, a focus on domestic politics is also problematic. Party control in Canada shifted—from the Conservatives to the Liberals and back to the Conservatives—with little effect on domestic climate policy.[1] Bernstein (2002) argues that international norms and

[1] Conservative Party Prime Minister, Brian Mulroney (1984–1993) led the government during the emergence of climate change as a political issue and through the UNFCCC negotiations. Liberal Party Prime Minister, Jean Chrétien's government (1993–2003) was responsible for the Kyoto negotiations and the response to the George W Bush administration's rejection of Kyoto in 2001. In 2003, following bitter divisions in the Liberal Party, Paul Martin assumed the role of prime minister and led the party to the creation of a minority government after the 2004 elections. Finally, Stephen Harper led the Conservative Party to a minority government after the 2006 elections.

the ideational incentives flowing from those norms are a significant force in shaping Canada's response. He claims that the convergence of international climate norms, domestic environmental norms and the Canadian government's concern for its international identity led it to pursue an international position that was not fully consistent with Canada's material interests (224–225). This argument raises important questions in the case of Australia as well. As noted above, the material and ideational forces acting on Australia's government were very similar to those of Canada, but they produced a very different outcome—at least in terms of Australia's response to Kyoto. Domestic political forces—particularly the federal power structure and the importance of western energy producing states—were obviously important in shaping national responses to climate change, but they cannot fully explain the foreign policy positions.

Canada has used its history of active involvement in international environmental affairs to bolster its reputation as a good international citizen. Canada was a leader in pressing for regulations to address ocean pollution in the 1970s (Kirton and Munton 1987). It played a significant role in the Law of the Sea negotiations and the development of the United Nations Conference on the Human Environment in Stockholm (Riddell-Dixon 1989; Cooper et al 1993). It was an advocate for international action on acid rain and ozone depletion. Individual Canadian diplomats and scientists have also played critical roles in the World Climate Programme and the development of the United Nations Environment Programme (Bernstein 2002). The Canadian government has thus cultivated an identity as an environmentally progressive state. The government's reputation as a good international citizen has emerged despite its difficulties in fulfilling the commitments that it supported. As Bernstein notes,

> Although Canada may often lead in identifying problems and supporting multilateral solutions, it is frequently unable to follow through with deep commitments owing to domestic constraints, especially in policy areas such as forests, biodiversity, and climate change that require substantial implementation across domestic sectors. (2002, 213)

This performance is consistent with Canada's use of symbolic politics to maintain its international identity while recognizing the political difficulty of fulfilling its obligations.

International leadership can both bolster a state's identity as a good citizen as well as expose the leadership to close scrutiny of its domestic performance. Cooper et al note, 'International activism is, thus, a two-edged sword. While it might well serve the international reputation of Canada and, by extension, its leaders, it also serves to intensify the process of domestic scrutiny' (1993, 162). For example, the Canadian government's championing of international action to address acid rain created close and unflattering scrutiny of its own pattern of industrial emissions. Its failure to reduce GHG emissions has also drawn attention. However, the fact that it continues to support international action and pledges to eventually achieve its domestic commitments has largely insulated Canada from the worst of the international opprobrium that has been directed at the US and Australia.

Canadian leadership on climate change emerged as a natural outgrowth of its efforts to address acid rain and ozone depletion. Canada was influential in the early framing of climate change as a political problem. Several Canadian diplomats and scientists were involved in the climate conferences of the mid-1980s (Bernstein 2002, 219).

Canada hosted the Toronto Conference on the Changing Atmosphere in 1988, which proved to be a turning point in convincing states to accept emission reduction commitments. Canada also supported an early beginning to climate negotiations—despite the strong objection of the US—as well as the inclusion of an emission reduction commitment in the UNFCCC. Canada was thus at the forefront of efforts to address climate change and was widely viewed at home and abroad as a leading state in the diplomatic response. Its role in the climate negotiations provided Canada with an opportunity to distinguish itself in the international system and promote an agenda distinct from that of the US.

After the completion of the UNFCCC, Canada began to face the political difficulties of meeting its commitment to reduce GHG emissions. Its domestic programmes have been largely voluntary and are unlikely to produce sufficient emission reductions.[2] Climate change split the country between its resource-dependent west and the more pro-environmental east. Canada joined a loose grouping of states called the JUSCANZ (pronounced 'juice cans')—Japan, the US, Canada, Australia and New Zealand—to press for differentiation in national targets and the inclusion of flexibility mechanisms, such as emission trading to reduce the costs of emission reduction. Canada could hide behind American opposition to binding emission reduction credits and could continue to support a stabilization of emissions at 1990 levels by 2010, secure in the knowledge that the US would block such a commitment.

The Kyoto negotiations placed Canada in a difficult position. In October 1997 the Clinton administration announced that it would accept a commitment to return American GHG emissions to 1990 levels by 2010. To maintain its leadership position, Jean Chrétien's Liberal government moved to beat the American commitment. At Kyoto, the Canadian delegation announced that it would commit to a three per cent emission reduction commitment and highlighted its position as a compromise between the American and European commitments.[3] This proposal exceeded what the government, and particularly the western provinces, had considered possible. When the US agreed at Kyoto to accept a seven per cent emission reduction commitment, Canada was forced into accepting a six per cent emission reduction commitment to avoid being labelled a laggard and being unfavourably compared to the US (Bernstein 2002, 222–223). Harrison quotes a Canadian official as asking rhetorically, 'Being good Canadians, can you imagine us letting the Americans get too far ahead of us' (2007, 102)? This commitment is

---

[2] The Conservative government's 1990 'Draft National Action Strategy on Global Warming' established the emphasis on public and corporate education, voluntary agreements, and subsidies rather than regulation and taxation as the focus of climate policy. The subsequent policy programmes, including 'Action Plan 2000 on Climate Change', the 2002 'Climate Change Plan for Canada' and the 2005 'Moving Forward on Climate Change: A Plan for Honoring our Kyoto Commitment' continued this approach—and the trend in Canadian emissions continued its upward path. The 2005 plan contained a more aggressive approach to reducing emissions but the new Conservative Party's 'Turning the Corner' programme has eliminated many of these provisions. The rules for the new programme are still being developed but there is significant debate surrounding the likely effectiveness of the carbon trading system proposed in the plan.

[3] Harrison argues that Canada acted at the urging of the US to bridge the gap with the European Union; regardless, Canada increased its commitment without an analysis of how the additional reductions could be achieved (2007, 101).

an excellent example of symbolic politics. The Canadian government had not altered its analysis of what was politically possible domestically. It took on a commitment that it knew was extremely unlikely to be met. However, the identity costs of undertaking a lesser commitment were too great.

Following the completion of the Kyoto Protocol, Canada continued to play a significant role in shaping the emerging climate regime. It argued, along with the US, for maximum use of flexibility mechanisms to achieve the Kyoto commitments as well as for maximum credits for carbon sinks. Canada received a number of concessions after the US rejected Kyoto, and it chose to ratify Kyoto without American participation. The Canadian position provided a stark symbolic act to differentiate Canadian support for multilateral action from American intransigence. Yet Canada remained unable to meet its Kyoto commitments without massive investments in the use of the flexibility mechanisms. Canada's support for Kyoto continued to be largely symbolic, as was clear in the failure to produce a concrete plan to achieve the Kyoto commitments (Harrison 2007, 107–110). However, the policy positions succeeded in solidifying a Canadian identity as a supporter of multilateral action to address environmental problems and helped Canada avoid international condemnation.

The approach of the end of the 2008–2012 Kyoto commitment period has made the Canadian position increasingly untenable. Canadian emissions have continued to rise, and it is gradually becoming apparent that there is little realistic chance of meeting the commitment. Acknowledging this fact in a 2007 speech at the Group of Eight Summit, Conservative Prime Minister Stephen Harper noted that,

> Normally, Canada is a country that prides itself on living up to its international obligations and commitments. But frankly, up to now, our country has been engaged in a lot of 'talking the talk' but not 'walking the walk' when it has come to greenhouse gases. (Harper 2007)

Harper went on to blame his Liberal predecessors for the failure to meet the target. He continued to emphasize the importance of addressing climate change and highlighted his government's 'Turning the Corner' initiative. The initiative lacks absolute emission reduction targets, and it instead focuses on reducing the carbon intensity of the Canadian economy. It is unclear whether this approach will produce substantial absolute emission reductions or whether it is an additional example of symbolic politics.

Australian climate policy

It could be argued that domestic party politics can explain Australia's response to climate change. Significant changes in Australia's international climate policy positions have followed changes in the party forming the national government.[4]

---

[4] Labor Prime Minister Bob Hawke (1983–1991) led Australia through the rise of climate change as a political issue, and his successor in the Labor Party, Paul Keating (1991–1996), led the government through the UNFCCC negotiations and into the initial phases of the Kyoto negotiations. Following the 1996 Liberal/National coalition victory, John Howard (1996–2007) led Australia through the Kyoto negotiations, the US rejection of Kyoto in 2001, and into the post-Kyoto negotiations. Finally, Kevin Rudd's 2007 Labor victory has produced a re-evaluation of Australia's climate policy in light of growing public support for action and international condemnation of Australian policy.

However, the foreign policy positions adopted by the various governments have been frequently undertaken for largely symbolic reasons: to manage Australia's identity before international and domestic audiences. Domestic policy has not varied significantly across parties; rather, the government has utilized international climate policy positions as convenient instruments to project clear differences with prior governments and to manage Australia's international image without substantial changes in domestic policies.[5]

The Australian government has frequently utilized international environmental policy as a tool for managing its international and domestic identity (Lightfoot 2006). The government effectively used its support of a mining ban in Antarctica—in spite of the strong opposition of the US and most other major actors with an interest in Antarctica—to bolster the Hawke Labor government's green credentials (Evans 1990; Cooper et al 1993). Australia used environmental issues of interest to South Pacific countries such as nuclear waste dumping, atmospheric nuclear testing and drift-net fishing to improve regional cooperation between Australia and its neighbours (Evans 1990). The government has also actively opposed Japanese whaling near Australia and Antarctica. These issues were easier to promote because they were largely consistent with Australia's material interests, but the government highlighted the symbolic rationale over the material interests in supporting these positions to solidify Australia's identity as a good international citizen.

The focus on maintaining an identity as a good international citizen was apparent in the public statements of many of Australia's political leaders, particularly under the Hawke and Keating Labor governments (1983–1996). For example, Gareth Evans, former Australian Minister for Foreign Affairs and Trade, in discussing the role of environmental affairs in Australia's foreign policy, noted that:

> I think it is worth emphasizing at the outset that the approach we take to international environmental issues is not one pursued in isolation. It is an integral part of the broader foreign policy interest we have in being—and being seen to be—a good international citizen. I have noted on several occasions that the concept of good international citizenship is not the foreign policy equivalent of Boy Scout good deeds. It is the logical consequence of Australia's place in the world where increasing interdependence makes global cooperation not some idealistic indulgence, but a pressing necessity. Because we recognize the interdependence of the world and because we need to have a say in how we are to solve global

[5] As in the Canadian case, Australian domestic climate policy has emphasized voluntary measures and subsidies rather than taxes and regulation. The 1992 National Greenhouse Response Strategy outlined a range of voluntary measures to encourage reductions in GHGs, but it proved ineffective. In 1995, the Keating Labor government launched its Greenhouse 21C initiative, which again emphasized voluntary measures. In 1998, after the Howard Liberal/National government had negotiated aggressively to limit its Kyoto commitment, it announced the creation of the Greenhouse Office and launched a National Greenhouse Strategy, which again emphasized voluntary measures and lacked emission reduction targets. Several Australian states have taken more aggressive actions, but Australian domestic policy has been remarkably consistent despite the changes in party control. The new Rudd Labor government has pledged to be much more aggressive, but it is too early to tell how effective it will be. For a good review and analysis of Australia's climate policy, see Christoff (2005).

> problems, we have placed, and will continue to place, a considerable emphasis in our foreign policy on multilateral diplomacy. This is particularly important at this time of extraordinary fluidity in international relations, which does create rather more opportunities for successful multilateral cooperation in areas such as the environment than have previously existed. (1990, 300)

In pursuing their internationalist and multilateralist approaches to Australian foreign policy, it would have been natural for the Hawke and Keating governments to support the climate negotiations. The Hawke government did not seek out a leadership position on climate change during the 1980s. However, during the late 1980s a norm began to emerge that all responsible members of the international system should adopt an emission reduction commitment (Cass 2006, 54–57). The Labor government duly responded in October 1990 by declaring that it would reduce Australia's GHG emissions by 20 per cent (based on 1988 levels) by 2005. This declaration was in line with other state commitments at the time and is an excellent example of symbolic politics. Australia sought to align itself with other progressive members of the international system and highlighted its position as a responsible state seeking to address a common environmental problem through a multilateral solution. However, this commitment was undertaken with little analysis of how it could be met. The Labor governments avoided the international opprobrium directed at the US for failing to accept an emission reduction target and were able to maintain Australia's identity in the international system during the period of transition after the end of the Cold War.

After the UNFCCC negotiations, the Keating government had become increasingly concerned by the potential costs of meeting a significant emission reduction commitment. The government continued to engage in international negotiations but retreated from its 20 per cent commitment. In 1996, it also launched a diplomatic effort to promote differentiation in national emission reduction commitments rather than common emission reduction targets (Papadakis 2002). This shift resulted in wide condemnation of the Australian position. Labor initiated the retreat from the emission reduction commitment, but the 1996 election of the new Liberal/National coalition government under John Howard dramatically changed the tone of Australia's negotiating position. The new Howard government emphasized the pursuit of Australian interests and the rejection of multilateral agreements pursued for the sake of appeasing international partners and maintaining Australia's reputation as a good international citizen. As the 1997 Australian White Paper on Foreign and Trade Policy put it, Australia is determined to 'resist the imposition of solutions to environmental problems which are contrary to Australia's national interest', and the government will 'stand aside from an international [environmental] agreement if ... others cannot be convinced that Australia's proposals are superior in terms of both the environment and the global economy' (Elliott 2001, 269). Climate change offered an immediate opportunity to demonstrate the government's new commitment.

The Howard government played to a domestic audience and raised the alarm that even a commitment to stabilize emissions at 1990 levels would result in the loss of 90,000 jobs. The government hardened its opposition to emission reduction commitments in the Kyoto negotiations and succeeded in securing a target that would permit Australian emissions to continue to grow. As Prime Minister Howard noted, 'We fought for Australian jobs every inch of the way and the outcome vindicates the government's stance' (Lunn and Garran 1997, 1). Australia's identity as a responsible international citizen was hurt by the aggressive diplomatic manoeuvring. However, the aggressive stand solidified the Howard government's reputation at home for standing up for Australia's interests in international affairs, even as it was condemned for its failings on climate policy (Christoff 2005, 33–34).

The Howard government sought to clearly differentiate its vision of Australia as an international actor from that of its predecessors, and climate change provided an immediate and high profile opportunity to do this. The government could have pursued the international path of least resistance and agreed to a small emission reduction commitment at the Kyoto negotiations, but instead it publicly emphasized its hard bargaining to secure an emission increase and lauded its 'victory' in the negotiations. The position taken by the government at Kyoto was largely consistent with Australia's economic interests, but the way that it was pursued was highly symbolic and directed at Australia's domestic audience.

Australia's later decision to walk away from its 'victory' at Kyoto was also consistent with Australia's economic interests, but it too was largely a symbolic act. Australia could have technically met its Kyoto commitment through the use of the flexibility mechanisms and various loopholes related to land-use negotiated by Australia, Canada, Japan and Russia after the US removed itself from Kyoto (Crowley 2007, 120). It could also have ratified the protocol and then failed to meet its commitment—as Canada has done—but it chose to align itself with the Bush administration and the US, which was consistent with the Howard government's emphasis on a close security and trading relationship with the US. The government even championed a range of bilateral and multilateral climate programmes—such as the Australian-US Climate Action Partnership and the Asia Pacific Partnership on Clean Development and Climate—as alternative international approaches to the Kyoto framework. Interestingly, the government pledged to meet the Kyoto commitment even as it refused to ratify the Protocol; thus, the change in its international position did not significantly affect its domestic policy (Papadakis 2002, 273). Crowley (2007) has called this Australia's attempt at 'faking' its support for addressing climate change. The government aligned its international positions with the US to manage its international identity even as it sought to placate domestic supporters of climate policy with commitments to meet its rejected Kyoto commitments.

Following the Labor victory in the 2007 elections, the new government under Kevin Rudd has also used climate policy as a symbolic force to demonstrate a break with the prior government. In a highly symbolic act, Prime Minister Rudd signed the ratification instrument for the Kyoto Protocol in December 2007. It was not immediately clear how Australia would meet its commitment, but the ratification was an unequivocal message to both international and domestic audiences that the nature of Australian foreign policy was now fundamentally different. Rudd's action was consistent with growing domestic support for efforts to tackle climate change; but, as in the Canadian case, more time will be necessary to determine whether the proposed policies will result in emission reductions or offer more symbolic action.

A focus on purely domestic political forces does not fully explain Australian climate policy. It is particularly important to note that the shifting international positions did not produce significant changes in domestic policies. The international positions were frequently changed symbolically to pursue other objectives that were, in some cases, unrelated to climate change.

Conclusion

Material forces and domestic politics are critical to understanding Canadian and Australian climate policy, but attempts by both countries to manage their national identities at the domestic and international levels have also been important. Canada and Australia utilized climate policy positions as symbolic signalling mechanisms to communicate with their international partners and domestic publics. Many of the positions adopted, and the manner in which those positions were presented, reflected symbolic politics. There were very few significant alterations in domestic policies that resulted from shifts in international positions, though this may change as the end of the Kyoto commitment period approaches.

There are many examples of the symbolism of climate policy. Canada's increase in its Kyoto emission reduction commitment from three to six per cent was not associated with a concomitant shift in domestic policies to achieve the additional reductions. Its ratification of Kyoto was not greeted with dramatic new policies to reduce emissions, though new policies are being widely debated today. Canada's policy has been driven in large part by a desire to differentiate itself from the positions adopted by the US, to signal its acceptance of the dominant international norms associated with climate policy and maintain its identity as a good international citizen. Canadian governments have faced significant criticism for failure to act more aggressively to meet the Kyoto commitments, but they have not had to withstand the level of condemnation directed at the US and Australia. Nevertheless, Canada will increasingly find it difficult to reconcile its rhetorical support for climate policy with its domestic failures to reduce GHG emissions. Thus far, it has not faced significant consequences for its failure to act. The question is whether the Canadian government will eventually pursue more aggressive action, or whether it can continue to affirm its international obligations while failing to fulfil them. This is a critical issue because it speaks to the question of norm compliance and the effects of international pressure on middle states in particular. If norm affirmation over a long period of time produces changes in behaviour, then there is hope that the promotion of preferred norms can eventually alter behaviour in a way to achieve the norm's desired consequences. However, if it is possible to continue to rhetorically affirm emergent norms without feeling compelled to fulfil associated policy obligations, then the political energy being put into shaming and coercion may be misplaced.

In the Australian case, the Hawke and Keating governments emphasized Australia's support for international action on climate change and affirmed the emerging norms related to climate policy, but the associated domestic policy changes were superficial. The government supported international action in part because—like the Canadian government—it was a way to signal Australia's support for a multilateral, rule-based international system and to bolster Australia's identity as a good international citizen. The Howard government's decision to take an international stand and reject Kyoto—thus demonstrating its bargaining prowess as an international actor—was largely consistent with Australia's economic interests; but the manner in which it was pursued was also driven by the desire to signal an identity to Australian society that was diametrically opposed to the international identity that the Hawke/Keating governments had pursued. Australia under the Howard government would be dedicated to the pursuit of Australian interests and not be bullied into accepting any commitment that was not consistent with those interests. Being a 'good citizen' was not a reason to undertake costly international commitments. The rejection of Kyoto, despite the Australian 'victory' over its negotiating partners at Kyoto and the promotion of alternatives to Kyoto, was also symbolic of Australia's alignment with the US and signalled the importance Australia attached to US-Australian relations. The new Rudd government's decision to make the ratification of Kyoto its first official act was similarly highly symbolic and was designed to signal a dramatic shift in Australian foreign policy and government priorities. The larger question is whether this change will translate into tangible domestic policies to achieve the Kyoto commitment. The history of Canadian and Australian domestic climate policy does not bode well, but this may be changing in light of growing public support and the approach of the end of the Kyoto commitment period.

Canadian and Australian climate policies provide important case studies of the effects of symbolic politics. Because support for sustainable development and protection of vulnerable populations has become increasingly associated with what is required of legitimate liberal states operating within the international system, environmental policy positions have become useful signalling tools for governments. In this sense, it is similar to human rights policy. However, because there are very limited enforcement mechanisms in the environmental and human rights areas, national leaders can affirm emergent international norms without being immediately forced to change domestic practices. Whether Canada and Australia are eventually forced to fulfil their commitments should provide some guidance in answering some of the questions raised in this analysis. Under what conditions will states fulfil international environmental commitments? What forces can successfully bring about shifts in state behaviour? Is international or domestic shaming sufficient to achieve anything beyond norm affirmation? Do international norms eventually force shifts in the positions of domestic interest groups sufficient to produce changes in state behaviour? These are some of the questions that need to be further addressed in these cases and others.

References

Bernstein, Steven (2002) 'International institutions and the framing of domestic policies: the Kyoto Protocol and Canada's response to climate change', *Policy Sciences*, 35:2, 203–236

Blühdorn, Ingolfur (2007) 'Sustaining the unsustainable: symbolic politics and the politics of simulation', *Environmental Politics*, 16:2, 251–275

Böhringer, Christoph and Carsten Vogt (2004) 'The dismantling of a breakthrough: the Kyoto Protocol as symbolic policy', *European Journal of Political Economy*, 20:3, 597–619

Cass, Loren R (2006) *The failures of American and European climate policy: international norms, domestic politics, and unachievable commitments* (Albany: State University of New York Press)

Christoff, Peter (2005) 'Policy autism or double-edged dismissiveness? Australia's climate policy under the Howard government', *Global Change, Peace and Security*, 17:1, 29–44

Cooper, Andrew Fenton (1997) *Niche diplomacy: middle powers after the Cold War* (London: Macmillan)

Cooper, Andrew, Richard A Higgott and Kim R Nossal (1993) *Relocating middle powers: Australia and Canada in a changing world order* (Vancouver: University of British Columbia Press)

Crowley, Kate (2007) 'Is Australia faking it? The Kyoto Protocol and the greenhouse policy challenge', *Global Environmental Politics*, 7:4, 118–139

Devetak, Richard and Jacqui True (2006) 'Diplomatic divergence in the antipodes: globalisation, foreign policy and state identity in Australia and New Zealand', *Australian Journal of Political Science*, 41:2, 241–256

Diez, Thomas (2005) 'Constructing the self and changing others: reconsidering "normative power Europe"', *Millennium: Journal of International Studies*, 33:3, 616–636

Edelman, Murray J (1964) *The symbolic uses of politics* (Urbana: University of Illinois Press)

Elliott, Lorraine (2001) 'Friends, allies or collaborators: environmental policy in the US-Australian relationship', *Australian Journal of International Affairs*, 55:2, 261–273

Evans, Gareth J (1990) 'Australia's foreign policy and the environment', *The Round Table: The Commonwealth Journal of International Affairs*, 315:1, 297–306

Finnemore, Martha and Kathryn Sikkink (1998) 'International norm dynamics and political change', *International Organization*, 52:4, 887–917

Harper, Stephen (2007) 'Prime Minister Stephen Harper calls for international consensus on climate change', Office of the Prime Minister, < http://pm.gc.ca/eng/media.asp?id = 1681 > , accessed 8 June 2008

Harries, Owen (2003) 'Punching above our weight?', *Boyer Lecture 6*, (Australia: ABC Radio National) 21 December

Harrison, Kathryn (2007) 'The road not taken: climate change policy in Canada and the United States', *Global Environmental Politics*, 7:4, 92–117

Holbraad, Carsten (1984) *Middle powers in international politics* (New York: St Martin's Press)

Howard, John (2001) 'Australia's international relations', Australian embassy at Pohnpei, < http://www.australianembassy.fm/interrelations.htm > , accessed 8 June 2008

Jones, David Martin and Andrea Benvenuti (2006) 'Tradition, myth and the dilemma of Australian foreign policy', *Australian Journal of International Affairs*, 60:1, 103–124

Kanie, Norichika (2007) 'Middle power leadership in the climate change negotiations: foreign policy of the Netherlands' in Paul G Harris (ed) *Europe and global climate change: politics, foreign policy and regional cooperation* (Cheltenham, United Kingdom: Edward Elgar), 87–112

Katzenstein, Peter J (1996) 'Introduction: alternative perspectives on national security' in Peter J Katzenstein (ed) *The culture of national security: norms and identity in world politics* (New York: Columbia University Press), 1–32

Kelly, Paul (2003) 'All the world's a stage', *The Weekend Australian*, 5–6 July, 17

Kirton, John and Don Munton (1987) 'The *Manhattan* voyages and their aftermath' in Franklyn Griffiths (ed) *Politics of the Northwest Passage* (Kingston: McGill-Queen's University Press), 67–97

Lightfoot, Simon (2006) 'A good international citizen? Australia at the world summit on sustainable development', *Australian Journal of International Affairs*, 60:3, 457–471

Lunn, Stephen and Robert Garran (1997) 'Our 1.42am greenhouse coup', *The Australian*, 12 December, 1

Manners, Ian (2002) 'Normative power Europe: a contradiction in terms?', *Journal of Common Market Studies*, 40:2, 235–258

Matten, Dirk (2003) 'Symbolic politics in environmental regulation: Corporate strategic responses', *Business Strategy and the Environment*, 12:4, 215–226

Meadows, Donella (1972) *The limits to growth* (New York: Universe Books)

Newig, Jens (2007) 'Symbolic environmental legislation and societal self-deception', *Environmental Politics*, 16:2, 276–296

Papadakis, Elim (2002) 'Global environmental diplomacy: Australia's stances on global warming', *Australian Journal of International Affairs*, 56:2, 265–277

Payne, Rodger A (2001) 'Persuasion, frames and norm construction', *European Journal of International Relations*, 7:1, 37–61

Reus-Smit, Christian (2002) 'Lost at sea: Australia in the turbulence of world politics', Working paper 2002/4, Australian National University, Canberra
Riddell-Dixon, Elizabeth M (1989) *Canada and the international seabed: domestic determinants and external constraints* (Kingston: McGill-Queen's University Press)
Rolfe, Mark (1997) 'The promise and threat of America in Australian politics', *Australian Journal of Political Science*, 32:2, 187–205
Sikkink, Kathryn (1998) 'Transnational politics, international relations theory, and human rights', *PS: Political Science and Politics*, 31:3, 517–524
Wood, Bernard (1988) *The middle powers and the general interest* (Ottawa: North-South Institute)
World Commission on Environment and Development (1987) *Our common future* (Oxford: Oxford University Press)

The climate regime and domestic politics: the case of Russia

Liliana B Andonova[1]

Introduction

Churchill spoke to a beleaguered Britain at the dawn of the Second World War, 'I cannot forecast to you the action of Russia. It is a riddle wrapped in a mystery inside an enigma; but perhaps there is a key. That key is Russian national interest' (Churchill 1939). Decades later, and despite the end of the Soviet Union, Russia's fluctuating stance on climate cooperation under the United Nations Framework Convention on Climate Change (UNFCCC)[2] seems wrapped in unpredictability. Russia initially opposed quantitative restrictions of greenhouse gas (GHG) emissions under the Climate Change Convention. The Kyoto Protocol, adopted in 1997, gave the country a significant economic stake in joining the climate regime by allowing it to increase its GHG emissions to the 1990 level or sell any unused

[1] The author is grateful to Colby College and to the Jean Monnet Fellowship Programme of the European University Institute for supporting the research presented in this article, as well as to Jayadev Vadakkanmarveettil for research assistance, comments and contribution.

[2] The term 'Climate Change Convention' is used hereafter for the sake of brevity and clarity.

amounts. Despite these favourable conditions, former President Vladimir Putin wavered for years before endorsing ratification of the Kyoto Protocol in 2004.

To unravel the shifting calculus of Russia's national interest in climate cooperation, this article examines the domestic impacts of the Climate Change Convention and the Kyoto Protocol. The article demonstrates that the international regime has helped change Russia's policy towards implementation of the Kyoto Protocol not just by altering the aggregated costs and benefits for the country, but also indirectly by creating a domestic political base in support of the treaty. A focus on Russia and the interplay of international and domestic climate politics is important from both a theoretical and policy perspective. From a theoretical perspective—although there is a vast literature on national policies and intergovernmental cooperation—few studies address the 'second image reversed'[3] question about the effects of the international climate regime on interests, coalitions, norms and policy-relevant knowledge within the state. Russia represents an ideal case for assessing alternative theories of regime impact in a domestic context, given its early scepticism about climate change, its reliance on fossil fuel exports and its insistence on unrestrained economic growth. The case allows one to examine which international mechanisms played a role in creating support for climate cooperation in a country characterized by low environmental concern and high reliance on fossil fuels. From a policy perspective, Russia remains a key global player with the third highest level of GHG emissions after the United States (US) and China. There are furthermore important similarities between Russia's primary interest in economic development and the interests of large, emerging markets such as China and India. Understanding the interplay between international and domestic climate politics in Russia can provide broader lessons about the conditions under which Russia and other rapidly growing economies may undertake obligations to strengthen climate cooperation.

The article proceeds as follows. The theoretical section conceptualizes three mechanisms of regime influence in domestic politics—socialization, capacity building and redistribution of resources—and specifies their observable implications. The impact (or lack thereof) of these mechanisms is then traced systematically in Russia's climate politics to explain the evolution of its climate policy. The article concludes by discussing the theoretical and policy implications of the analysis.

Domestic impacts of international regimes

Early debates about international regimes focused on whether and how institutions influence intergovernmental relations and cooperation.[4] Increasingly and particularly in the study of global environmental politics, scholars have turned their attention to questions of regime implementation, compliance and effectiveness.[5] Much of the regime literature, however, takes states as the main unit of analysis. Constructivist approaches illuminate the influence of international suasion and norms on state behaviour. Rational choice studies

[3] This is a term coined by Gourevitch (1978).
[4] For a review see Simmons and Martin (2002).
[5] For an overview of issues and methods see Miles et al (2002).

emphasize the role of issue linkage, monitoring, resource transfer and iterative-bargaining in facilitating cooperation. These theoretical approaches imply distinct mechanisms of domestic influence, but their interplay with domestic politics has rarely been the focus of research, leaving a persistent gap in the literature (Botcheva and Martin 2001; Schreurs and Economy 1997; Simmons and Martin 2002). By examining the domestic impacts of the climate regime, the article contributes to addressing this theoretical challenge.

The main dependent variables the study is concerned with are domestic political processes, rather than aggregate state behaviour or policies. Building on theories of international institutions, compliance, and political economy, the analysis distinguishes between three types of regime influence on domestic politics: socialization, capacity building and redistribution of resources.

Socialization is emphasized by constructivist studies, which point to the role of international regimes in validating and diffusing consensual knowledge and norms in ways that affect states' perception of interests and 'appropriate' behaviour (Checkel 2001; Haas 1989; Keck and Sikkink 1998). These socialization mechanisms, however, are often processed through domestic politics rather than simply through state-to-state interaction. The diffusion of consensual knowledge and information, for example, depends critically on the strength of domestic epistemic communities (Haas 1989). The diffusion of norms, in turn, depends on advocacy organizations and their ability to build transnational coalitions and exert pressure on governments (Keck and Sikkink 1998). We should therefore observe a regime-induced strengthening of epistemic and advocacy communities and of their ability to influence policymaking if socialization mechanisms affect domestic politics.

Capacity building is emphasized in studies of compliance with international law. Whereas monitoring and enforcement seek to prevent or punish defection, softer management mechanisms strengthen the capacity of states to mitigate involuntary noncompliance (Brown Weiss and Jacobson 1998; Tallberg 2002). Financial transfers, training and assistance for institution building and reporting are particularly important capacity building mechanisms in environmental regimes due to the scientific complexity of agreements and financing required for implementation. Such interventions are typically viewed as mechanisms to alter state behaviour, but they can have important impacts on domestic politics and institutions. The provision of resources can affect the leverage of participating departments within the executive bureaucracy and increase their leeway to pursue a specific policy agenda. International programmes often establish new focal points for policy-making. The executive branch can use cooperation to tie its hands behind international commitments and manoeuvre between international and domestic planes of negotiation to advance its agenda (Putnam 1988). One way in which we can tell that an international regime has influenced domestic politics through capacity building mechanisms is to trace observable changes in the position of participating agencies and their ability to promote a preferred set of policies.

International regimes also influence domestic politics through redistributing costs, benefits and risks across sectors of society. Political economy analyses in the 'second image reversed' tradition have made a significant contribution by illuminating the redistributive implications and political consequences of the global trade regime (Gourevitch 1978; Milner 1988; Rogowski 1990). The domestic distributional effects of environmental agreements have not been as systematically analysed. The climate regime, in particular, is likely to have substantial and differential cost and benefits implications not only across countries, but also within countries. Sectors reliant on fossil fuels face higher costs associated with capping GHG emissions, whereas sectors vulnerable to climate variability, and those developing alternative energy technologies, stand to gain most from climate mitigation. Since the Kyoto Protocol explicitly links GHG reductions to markets and technology investment, it could facilitate unlikely coalitions between environmentalists, businesses interested in carbon markets and public entities, such as municipal or regional governments interested in benefiting from these linkages.

The domestic cost-benefit calculus with respect to the climate regime can be complicated by factors such as the potential for explicit or implicit subsidization of emission reduction projects, uncertainty with respect to the development of new technologies and emission trading markets, and the opportunities for companies to bundle multiple technologies, products and services. The redistributive mechanisms of regime influence imply observable changes in domestic interests and coalitions that can be traced to the rules of the regime. These changes could push politics towards supporting or opposing the regime, depending on the size of the coalition of regime winners, the political influence of blocking interests and the domestic institutional context.

Table 1. Typology of domestic impacts of international regimes

Type	*Mechanisms*	*Actors affected*	*Implied change in domestic politics*
Socialization	Consensual knowledge diffusion	Epistemic communities	Empowerment of epistemic communities supportive of regime Uptake of consensual knowledge in policy process Change of policy frames and perception of interests
	Norm diffusion	Advocacy organizations	Empowerment of advocacy organizations Changes in public opinion, policy discourse, framing
Capacity building	Financial and institutional assistance, training	Public administration	Increased political influence of participating agencies New focal points for policy-making, issue linkage
Redistribution	Costs and benefits of regulation	Interest groups	Change in domestic preferences with respect to international regimes
	Explicit/implicit subsidization		New intra- and cross-sectoral coalitions

Table 1 summarizes the three types of mechanisms of regime impact, the actors most likely to be impacted and their observable implications with respect to domestic politics.

Important analytical gains can be made from disaggregating and analysing the mechanisms of regime influence on domestic politics. This approach allows one to assess the explanatory power of alternative theories in the domestic context and to examine regime effects with greater empirical detail and confidence. It decreases reliance on counterfactuals and makes it easier to distinguish between changes in state behaviour in response to exogenous factors and those that can be directly linked to the regime through political processes. The 'second image reversed' perspective can also shed light on the unintended or unexpected consequences of the regime, and on the interactive effects of multiple mechanisms. Following a short background section, the article explores these dynamics in the context of Russia's climate politics.

Background: from sceptic to kingmaker in the climate regime

Climate change is a global common problem created by the accumulation of increasing amounts of GHGs in the atmosphere. Cooperation to address the issue has to overcome multiple collective action problems associated with the pervasive economic implications of regulating fossil fuels, incentives to free-ride, necessity to invest resources now for the benefit of future generations, and reliance on complex and evolving scientific understandings of earth systems. The justice implications of climate cooperation are also not trivial. Six of the largest economies in the world—the US, the European Union (EU), which now consists of 27 member states, China, Russia, Japan and India—account for approximately 60 per cent of the global emissions of the six major GHGs, making the rest of the world responsible for only about 40 per cent of global emissions (Figure 1). Among these

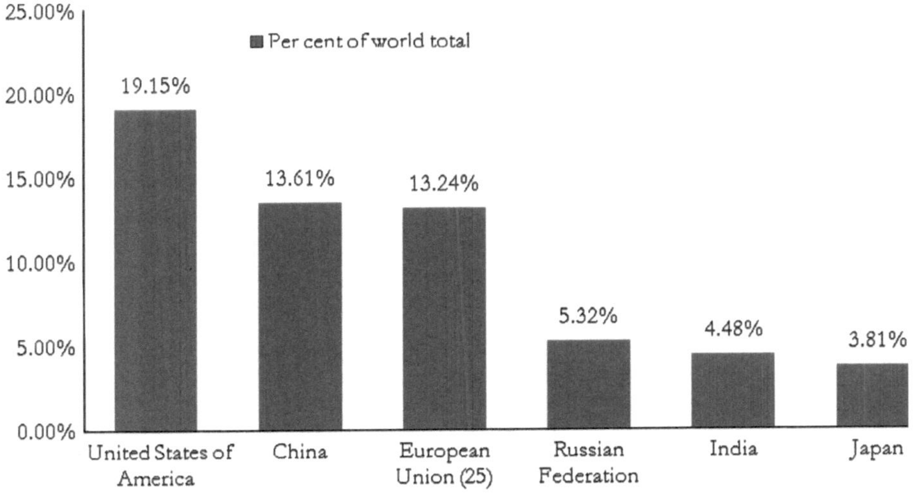

Figure 1. Top contributors to global greenhouse gas emissions. Total 2000 emission of CO_2, CH_4, N_2O, HFC, PFC and SF_6, as percentage of world total, http://cait.wri.org/, accessed 24 July 2008

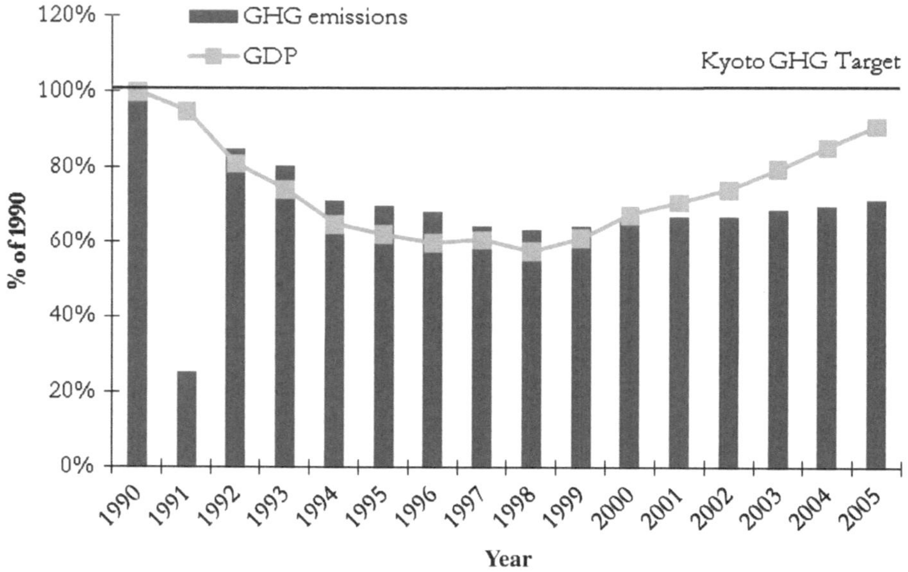

Figure 2. Gross domestic product (GDP) and GHG emissions (CO_2, CH_4, N_2O, PFC and SF_6) of Russia as a percentage of 1990 level

top contributors to climate change, some countries have enjoyed long periods of unrestrained fossil-fuel-based industrial development, whereas others are just emerging from poverty and growing rapidly. The negative impacts of climate change are furthermore likely to fall disproportionately on the least-developed countries.

In this constellation of global players, Russia is a particularly important case. It is an industrialized country and member of the Group of Eight (G8), with industrial production accounting in 2006 for 39 per cent of its Gross Domestic Product (GDP) and services accounting for another 56 per cent (World Bank 2007). However, Russia experienced a sharp economic downturn after the end of communism and through the 1990s, culminating in the financial crisis of 1998. As a consequence, it has prioritized rapid economic growth much like other emerging markets. The country has unmatched fossil fuel resources, controlling about 42 per cent of the world's natural gas resources, 32 per cent of brown coal stocks and 12 per cent of oil stocks (Russian Federation 2002, 28). Energy exports and rising fuel prices have stimulated a remarkable economic recovery since 1999 (Figure 2). At the same time, Russia remains one of the most energy-intensive economies in the world, using three to four times as much energy per unit of GDP than EU states (Grubb 2004) and one-and-a-half to two times as much energy per unit of GDP than Canada—a country with a similarly cold climate and vast territory.[6]

Russia's position has undergone an evolution from sceptic to a critical player in the climate regime, whose support for the Kyoto Protocol became a necessary

[6] Comparison based on GDP per unit of energy use statistics (constant 2005 Purchasing Power Parity [PPP] $ per kg of oil equivalent) as provided by World Bank (2007).

condition for the treaty's entry into force and implementation. At the 1992 negotiations on the Climate Change Convention, the Russian delegation questioned theories of anthropogenic influence on the climate system and maintained that global warming would largely be to its benefit. Keen on joining international institutions in the aftermath of communism, Russia was one of the first countries to ratify the Climate Change Convention, which established a general framework and principles for cooperation. Subsequently, however, it aligned with the members of the Organization of the Petroleum Exporting Countries (OPEC) in opposing any quantitative emission limits. Eventually Russia left the OPEC camp and at the 1997 conference of parties in Kyoto, sided with the US, Japan and Australia in pressing for more lenient targets for industrialized states than the 15 per cent reduction from 1990 levels proposed by the EU (Kotov 2002; Moe and Tangen 2000; Nikitina 2001).

At Kyoto, Russia wielded considerable bargaining power as a result of two factors. First, being one of the most significant industrialized emitters of GHGs, Russia's participation was essential for the regime, but domestic interest in climate cooperation was largely nonexistent. The country's economy and GHG emissions had shrunk by 39 per cent and 34 per cent respectively between 1990 and 1997 (Figure 2). The domestic priority of economic growth gave negotiators substantial bargaining leverage at the international table. Russia's position was that it would not commit to any target beyond stabilization at 1990 levels, as it was entitled to regain its pre-1990 economic strength (Kotov 2002; Kotov and Nikitina 2003). The second source of Russia's bargaining leverage at Kyoto was its position as a mediator between the EU and the US. Together with the US, Russia was in favour of the unrestricted use of the flexible economic mechanisms such as emissions trading and Joint Implementation (JI),[7] since it expected to be a net seller of emission reduction allowances due to its already large reductions of GHG emissions compared to the base year 1990.

Consequently, Russia achieved by far the most favourable terms of all industrialized states. The Kyoto Protocol establishes Russia's emissions cap at the 1990 level, effectively allowing it to increase its 1997 emissions by about 34 per cent (Figure 2). The agreement adopted three flexible instruments—emissions trading, JI and the Clean Development Mechanism. It thus institutionalized potentially substantial side payments for Russia in the form of excess emission reduction allowances (often called 'hot air') and opportunities for attracting carbon-reducing investment through JI. As the US commitment to the climate regime receded, Russia's bargaining power grew. To enter into force, the Kyoto Protocol required 55 ratifications and the accession of industrialized countries accounting for no less than 55 per cent of total industrialized country emissions, which made Russia's participation critical in the absence of the US. Russia got further concessions at the 2001 Marrakech Conference of Parties, which effectively increased its GHG emission allowance by 37 Metric Tons of Carbon (MtC) for GHG absorption from land use and forests (Grubb 2004).

---

[7] The mechanism, 'Joint Implementation' is defined in article 6 of the Kyoto Protocol. It allows a country to meet its emission reduction commitment by earning emission reduction units (ERUs) from an emission reduction or emission removal project in another signatory of the Protocol. For more information, see the UNFCCC website <http://unfccc.int/>.

Russia could expect virtually no cost of compliance with the Kyoto Protocol and significant potential benefits. Following a unitary actor logic, most countries and analysts expected rapid ratification in time for the 2002 World Summit on Sustainable Development. Yet Russia did not oblige. At the Summit and in subsequent years, Russia's officials spoke with multiple and conflicting voices (Henry and Sundstrom 2007). As if to confirm the unpredictability of Russia's position, President Putin only endorsed the treaty for ratification in September 2004, shortly after Russian officials had made doomsday statements about the imminent end of the Kyoto Protocol.[8] Post factum, Russia's ratification was interpreted as a tacit, tit-for-tat deal with the EU to secure the EU's support for Russia's membership in the World Trade Organization (WTO). Watching Russia from the outside as a unitary actor thus makes it difficult to predict at which point its national interest would align with the objectives of the climate treaty, and whether the Kyoto Protocol has influenced the formulation of the country's interest in important and lasting ways.

To ascertain the impact of the Kyoto Protocol on Russia's climate policies, it is necessary to situate the mechanisms of international influence in the country's institutional context. Russia has been described as a 'superpresidential regime' (Shevtsova 2000, 32). Russia's first post-Soviet President Boris Yeltsin consolidated decision-making power in the hands of the presidency with the proposal and adoption of a new constitution in 1993. Yeltsin governed in close relationship with oligarchic economic elites and a group of tight-knit advisors, creating a situation widely described as 'rent-seeking' and 'state-capture' (Yakovlev 2006, 1033–1034). The Federal Assembly, Russia's legislative body, imposes weak to nonexistent checks and balances on the President, who may nominate the Prime Minister, dissolve the Assembly, chair governmental meetings and initiate legislation (Easter 1997). The President is furthermore entrusted with guiding the country's foreign policy. International treaties can be ratified by the Federal Assembly only after the President, Prime Minister or First Deputy Prime Minister proposes the document for ratification. After ratification, the President's signature is still required for the treaty to enter into force.

Centralization of executive power continued under President Vladimir Putin. Putin, however, sought to distance himself from some of the legacies of his predecessor by seeking to strengthen the Russian state and restructure relations with business (Colton 2007; Yakovlev 2006). He selectively prosecuted oligarchs such as Vladimir Gusinsky, the owner of the NTV television network, and Mikhail Khodorkovskii, the owner of the Yukos oil company whose influence was deemed threatening to the President's political power. According to Yakovlev, a new relationship of 'equidistance' was gradually established between federal authorities and business elites, characterized by

> the emergence of giant conglomerate business groups that possessed sufficient influence to 'counterbalance' certain consolidated governmental bodies and their competition for power and resources. (Yakovlev 2006, 1046)

---

[8] See Myers and Revkin (2003); Federal News Service, Inc, 'Official Kremlin International News Broadcast on Friday, July 9, 2004', <http://www.sysecol.ethz.ch/Articles_Reports/Illarionov_Interv._9.Jul.04.pdf>, accessed 24 July 2008.

Putin also increased the share of state ownership and management of energy and other natural resources.

The authority of the federal bureaucracy was consolidated through reforms in the tax code, land code and the judiciary. Provincial governors were removed from the upper house of the Federal Assembly and, in 2005, the direct election of governors was replaced by a system of presidential nominations, whereby governors are only confirmed by the vote of the legislatures in their provinces. With respect to civil society, there has been a trend towards increased control over the media and efforts to curb international funding and activities of nongovernmental organizations (NGOs). The power of President Putin was further consolidated after the 2003 electoral victory of the pro-Putin United Russia party, which gained control in the lower house of the Federal Assembly, the Duma. Putin hand-picked his successor Dmitry Medvedev, who was elected President in March 2008. Medvedev subsequently appointed Putin as the new Prime Minister of Russia.

Given the high level of centralization of executive power in Russia, it might be tempting to assume that Russia's climate position was and will remain highly unpredictable because it reflects the changing preferences of a single, powerful leader. Such an assumption, however, would be an oversimplification of the political reality and would fail to consider the multiple and sometimes competing political interests and advisors who influence the formulation of national policies (Colton 2007). The analysis in the following sections therefore seeks to establish how the climate regime has affected actors with access to executive decision-making, what political processes have ensued and what their impact on climate policies is.

International socialization and the battle of ideas in the Kremlin

The diffusion of scientific knowledge and environmental norms has played a relatively limited role in influencing domestic discourse and politics in Russia. At first glance, this is surprising given the evolving international consensus on climate science and vulnerability of the northernmost regions of the hemisphere, where a large portion of Russia is situated. As discussed previously, the impact of socialization and knowledge diffusion depends critically on the ability of domestic and transnational epistemic and advocacy communities to internalize and interpret norms and information in ways that influence the decision-making process (Haas 1989; Keck and Sikkink 1998). The gate-keeping role of climate sceptics in the epistemic and policy establishment of Russia and the weak position of civil society organizations explain the limited influence of scientific information and normative action. By contrast, the diffusion of economic expertise on emission scenarios and on the net economic benefits implied by the Kyoto Protocol were instrumental in helping to shift policy debates away from the 'deficiencies' of the Kyoto Protocol towards a consensus on the economic rationale for supporting it.

In the long run-up to the ratification of the Kyoto Protocol, an outspoken climate science sceptic, Yuri Izrael, was the most influential scientific advisor to President Putin. Professor Izrael is an established Soviet climatologist and the Director of the Institute of Global Climate and Ecology under the Russian

Academy of Sciences and the Federal Service for Hydrometeorology and Environmental Monitoring (Roshydromet). Despite his participation in the International Panel on Climate Change, the main international scientific body of the Climate Change Convention, Professor Izrael publicly questioned the interaction between human-induced and nature-induced changes in the climate system, emphasizing the comparatively greater impacts of the latter. As recently as 2007 Izrael maintained: 'I think the panic over the global warming is totally unjustified. There is no serious threat to the climate.'[9] He has also been sharply dismissive of the Kyoto Protocol, arguing that it is 'scientifically ungrounded and economically hazardous to Russia and should be abandoned'.[10] In a less flamboyant fashion, other voices from Russia's scientific establishment have maintained a similar line of scientific and policy scepticism,[11] reflected also in an official 2004 study of the Russian Academy of Sciences, which emphasized that the Kyoto Protocol was not supported by scientific evidence and implied a significant economic risk for Russia (Avdeeva 2005, 293).

However the growing international consensus on climate change and societal vulnerability has not completely bypassed Russia. Russian scientists focusing on sectors such as hydrology, agriculture and permafrost areas have drawn attention to the vulnerability of these sectors (Nikitina 2001). The sensitivity of permafrost-covered territories, which account for 67 per cent of Russia's land mass, is a particularly salient topic of scientific research. These regions are both ecologically fragile and economically vulnerable to damages to infrastructure and human settlements, which are likely to be impacted by climate change (Anisimov and Reneva 2006; Russian Federation 2006). Furthermore, thawing permafrost in the Russian north could result in the rapid release of large quantities of methane and carbon, accelerating climate change and its associated risks (Anisimov and Reneva 2006).

Despite increasing scientific recognition of the vulnerability of permafrost areas and other sectors, a text analysis of Russia's National Communications to the Climate Change Convention submitted between 1995 and 2006 shows a continued tendency to emphasize first the potential for positive impacts of climate change before discussing risks. Both the 2002 and 2006 Communications begin the discussion of vulnerability by highlighting positive impacts, stating, for example: 'The observable warming of the climate, overall, is likely to be favourable for agriculture in Russia ....' (Russian Federation 2006, 79, translation by author). The sensitivity of the permafrost areas were only discussed in the later section of the vulnerability chapter which acknowledges that climate change 'may cause serious negative consequences for settlements and other managed territories' (Russian Federation 2002, 19, translation by author; see also Russian Federation 2006, 86–88).

Norm diffusion has had a similarly limited impact on Russia's climate politics due to the weak position of environmental groups and low public concern with

---

[9] Yuri Izrael for RIA Novosty, <http://en.rian.ru/analysis/20070418/63856919.html>, accessed 24 July 2008.

[10] Statement reported in <http://newsfromrussia.com/world/2004/07/17/55051.html>, accessed 1 August 2008.

[11] See 'Kyoto Protocol: Pros & Cons', <http://eng.globalaffairs.ru/books/548.html>, accessed 24 July 2008.

climate change. The most significant climate campaigns in Russia have been carried by transnational NGOs such as the World Wildlife Fund International, Greenpeace and the Climate Action Network Europe (Henry and Sundstrom 2007). The access of these organizations to top level decision-making has been limited due to Russia's highly centralized institutional structure, and particularly after the adoption in 2005 of the regulations restricting the activities of NGOs. Both the lack of a domestic lever and a hostile governmental apparatus have limited the ability of environmental advocacy groups to significantly influence domestic politics and discourse.

Two of the most prominent analysts of Russia's environmental policy in transition, Vladimir Kotov and Elena Nikitina (2003), point out that Russian NGOs remain largely of a 'decorative character', while the public remains undivided in its indifference to climate change as citizens are faced with more urgent economic problems. International public opinion surveys have shown a marked increase from approximately 43 per cent in 2003 to 58 per cent in 2005–2006 in the share of Russian respondents who consider climate change to be a very serious problem (Brewer 2007). However, a recent press release of Greenpeace Russia still emphasizes that: 'According to sociological surveys, more than half of Russians are unaware of the problem of climate change. Most believe that for frigid Russia warming would not be a misfortune, but a blessing.'[12]

Knowledge diffusion did play a considerably greater role in Russia in strengthening the policy consensus on the economic and technological benefits of the Kyoto Protocol. Although the Protocol allocated Russia a generous cap of excess emissions, the policy circles close to the President were still concerned about a number of issues, such as the absence of US participation, the uncertainty over Russia's economic benefits from the flexible mechanisms given the EU's reservation about 'hot air', and the possibility that the Kyoto cap might impose an unnecessary constraint on economic growth. These concerns opened up new space for rhetorical battles between high level supporters and opponents of Russia's accession to the Kyoto Protocol. Andrei Illarionov, Putin's chief economic advisor between 2000 and 2005, promoted the opinion that the emissions trading pie would be miniscule and dismissed the Kyoto Protocol in a rather flamboyant fashion, as

> one of the biggest, if not the biggest, international adventures based on man-hating totalitarian ideology which, incidentally, manifests itself in totalitarian actions and concrete events, particularly academic discussions, and which tries to defend itself using disinformation and falsified facts. ... It's unpleasant to say but I am afraid it's undeclared war against Russia, against the entire country, against the left and the right, against the liberals and the conservatives, against business and the Federal Security Service, against the young and the old who live in Moscow or in the provinces.[13]

---

[12] See <www.greenpeace.org/russia/en/press/releases/1612802>, accessed 24 July 2008.

[13] The Federal News Service, Inc, 'Official Kremlin International News Broadcast on Friday, July 9, 2004', <http://www.sysecol.ethz.ch/Articles_Reports/Illarionov_Interv._9.Jul.04.pdf>, accessed 24 July 2008.

Despite the statements of Illarionov appearing at times to be purely inflammatory rhetoric, they did pose bluntly the question of the potential costs and benefits associated with the Kyoto Protocol. In response to such concern, countries interested in the implementation of the climate regime exerted considerable effort in providing credible assessments of the potential benefits and institutional requirements associated with carbon-trading in Russia. The World Bank together with the governments of Finland and Switzerland supported a Joint Implementation Strategy Study for Russia as early as 1997–1999 (World Bank 1999). The study was prepared by a team of Russian experts, including staff from the Ministry of Economic Development and Trade, the former State Committee of Environmental Protection and the nonprofit Bureau of Economic Research. It produced emission forecasts and scenarios of the market potential for emissions trading and JI in Russia.

The European Commission also sponsored a range of conferences and analyses on the implementation of the Kyoto Protocol. The government of the United Kingdom (UK) and the Royal Institute for International Affairs[14] played a particularly important role in undertaking independent studies—conducted with the support of Russian experts—on the costs and benefits of the Kyoto Protocol and disseminating these findings through international workshops and networks. An article by Michael Grubb echoes the transnational economic consensus that was gradually established:

> Even with the most rapid economic growth there are no internally consistent scenarios in which Russian emissions breach its Kyoto allowance: Russia will be net seller, and the Kyoto mechanisms could help attract foreign capital linked to energy modernization. (Grubb 2004, unnumbered page)

Transnational projects for the diffusion of economic expertise had several impacts in Russia. Firstly, they helped the early establishment and strengthening of an economic expert community on the flexible mechanisms of the Kyoto Protocol within or close to the federal bureaucracy and decision-making apparatus in Russia. Secondly, projections for the potential of carbon-trading and JI were based on domestic expertise, official government data and growth forecasts, which increased the domestic salience and credibility of their conclusions. Thirdly, the economic expert communities found a receptive interlocutor and home base in the Ministry of Economic Development and Trade (Korppoo et al 2006; World Bank 1999).[15] This alignment contributed to the formation of a policy and expert network centred on the Ministry of Economic Development and Trade, which provided an important epistemic counterbalance to the sceptical scientific and economic advisors of the President. As Avdeeva observes:

> Quite a few Russian experts are convinced that setting mandatory limits on GHG emissions could contribute to reducing the energy intensity of Russia's GDP, promote technical renovation of the Russian petroleum industry and prompt

[14] These studies are now published by Chatham House, <http://www.chathamhouse.org.uk/>.

[15] Interview conducted on the basis of anonymity at the Director-General Environment office, European Commission, Brussels, 21 June 2005. See also World Bank (1999) and Korppoo et al (2006).

> greater diversification of its energy exports by raising the share of processed products. ... By different accounts, Russia could earn as much as $8 billion–$20 billion by selling its unused emissions under the Protocol on the world market. (Avdeeva 2005, 297–298)

The 2002 and 2006 National Communications to the Climate Change Convention similarly did not anticipate any scenario under which the country would exceed its 1990 level of GHGs emissions by 2008–2012. Even with the most optimistic growth scenario, emissions were forecasted to reach 79.7 per cent of 1990 level by 2010, leaving considerable potential for carbon-trading (Russian Federation 2006, 78). Over time, a network of economic and policy experts helped establish a policy consensus on the economic significance and limited risk associated with the Kyoto Protocol.

Capacity building and the realignment of bureaucratic politics

Russia has participated in multiple international projects related to the implementation of the Climate Change Convention and the flexible mechanisms of the Kyoto Protocol. International programmes have supported the preparation of GHG emission inventories at different levels (company, regional and national), training for project portfolio development and private sector engagement, and identification of regulatory and policy instruments necessary to facilitate the functioning of carbon markets. Bilateral donors—including Denmark, Germany, Japan, The Netherlands, Sweden, Switzerland, the UK and the EU—have generously financed capacity building cooperation. These governments have a strong interest in securing the entry into force of the Kyoto Protocol, and some are prepared to take advantage of the flexible mechanisms to meet their national commitments. This institutional support has interplayed with the restructuring of the Russian federal bureaucracy to facilitate a power shift in the management of climate polices towards the Ministry of Economic Development and Trade.

The Ministry of Economic Development and Trade did not play an important role in shaping Russia's position in the climate regime prior to the Kyoto agreement. Roshydromet was the lead agency in the Inter-Agency Commission on Climate Change, established in 1994, to coordinate climate policies (Kotov 2002). The Kyoto Protocol gradually came to be seen by top officials in the Ministry of Economic Development and Trade, such as the former Minister German Greff, as an opportunity to pursue economic objectives and expand the agency's resources and mandate. It was very much in the organizational interest of the Ministry to position itself as the main coordinating governmental body for carbon transactions. Such transactions could potentially provide a significant flow of resources through the organization and expand the turf of the Ministry. The Protocol's flexible instruments also gave the Ministry leverage to advance economic priorities such as reducing energy intensity, encouraging investment in technology modernization, and diversification away from reliance on oil and natural gas exports as the main sources of revenue (Avdeeva 2005). International projects contributed to this interest by placing an emphasis on the economic instruments of the Kyoto Protocol and engaging the Economic Ministry, often as a lead agency, in capacity building and economic assessment programmes

(Russian Federation 2006; World Bank 1999). The Ministry of Economic Development and Trade became widely regarded by international donors as their main domestic counterpart.[16]

As a consequence of its increasingly prominent role in climate policy, both the Ministry of Economic Development and Trade and Roshydromet became co-chairs of the Inter-Agency Commission on Climate Change and in 2002 the Ministry of Economic Development and Trade became the lead agency (Korppoo et al 2006). This development was critical in Russia's domestic politics. The original chair, Roshydromet, had weak political authority as its expertise pertained primarily to climate science and weather services (Nikitina 2001). The Ministry of Economic Development and Trade brought greater political clout to the work of the Commission, access to the top echelons of political power and redefined climate change as a policy to promote efficient economic development. It is not accidental that in 2004 the Ministry of Economic Development and Trade negotiated both the terms of EU's support for Russia's WTO accession and the tacit agreement to proceed with the ratification of the Kyoto Protocol. As the top executive negotiator on both issues, the agency was in a position to successfully manoeuvre between the domestic and international planes of negotiation and to link issues of economic development and environmental cooperation to advance two complementary foreign policy objectives.

Other institutional actors also developed an interest in the ratification and implementation of the Kyoto Protocol. The National Agency for Direct Investments—an organization with independent legal status but supported by the Ministry of Economic Development and Trade—and the Russian Chamber of Trade and Industry, participated in a range of workshops on JI organized in the context of the Climate Change Convention. By early 2004, the Agency presented itself as a vocal supporter of the Kyoto Protocol ratification, emphasizing the investment benefits of JI projects and the potential for decoupling economic and emissions growth (Agency for Direct Investments of the Russian Federation 2004). Regional authorities also prepared, with international support, GHG inventories and baselines to enable them to attract JI investments. Over a period of several years prior to the ratification of the Protocol, at least 35 of the 89 regions of Russia demonstrated a willingness to strengthen their climate policy capacity and participate in the Kyoto Protocol (Nikitina 2001; Karas 2004). The interest of regions and investment institutions was used by the Ministry of Economic Development and Trade to further boost political support for the Protocol (Russian Federation 2006).

Given the central role of the executive branch in Russia's policy-making, the shift in bureaucratic politics towards the influential Ministry of Economic Development and Trade was critical in swaying the government's decision in favour of the Kyoto Protocol. This shift was informed by the agency's own interest in the flexible mechanisms and was supported by a long chain of international projects encouraging their implementation, which helped place the Ministry at the centre of climate cooperation.

---

[16] Interview conducted on the basis of anonymity at the Director-General (DG) Environment office, European Commission, 21 June 2005.

Distributive impacts and the unlikely supporters of the Kyoto Protocol

One of the least anticipated impacts of the climate regime on Russian domestic politics was the turnaround of major energy companies into supporters of the Kyoto Protocol. The change in these actors' interests can be traced directly to the rules instituted by the treaty and their distributional implications. The excess emission quotas Russia gained were viewed by the international community as a substantial side payment for the country as a unitary actor. As it turned out, these economic incentives had to be processed first through domestic interest group politics and, as already discussed, through bureaucratic politics to ultimately influence the position of the country's top executive. The change in the interests of key energy companies in Russia is particularly significant because of the initial perception of the sector as a potential loser in climate cooperation, which informed the country's early alignment with the OPEC (Moe and Tangen 2000).

Russia's long-time electricity monopoly, the United Energy Systems (UES), exemplifies most prominently the change in domestic interests. Prior to the restructuring of the UES, which started in 2006 and was completed in June 2008 with the separation of 23 state and privately owned power generators, the company was one of the largest single utilities in the world. Its majority stake (52 per cent) was controlled by the government, and about 48 per cent of its shares were privately owned by foreign and Russian entities and individuals (Chubais 2001). UES provided 85 per cent of the electricity power generated in Russia, controlling 440 thermal and hydropower plants, 72 regional utilities and the country-wide grid (Chubais 2001). Russia's nine nuclear power plants are separately controlled by the Ministry of Atomic Energy. The thermal power generation of the company relied primarily on natural gas (70.6 per cent), coal (25.6 per cent), fuel oil (3 per cent) and other fuels (0.8 per cent) (Chubais 2001). UES accounted for as much as 30 per cent of Russia's GHG emissions (O'Brian 2004).

Despite its substantial reliance on fossil fuels, UES was one of the first actors to voluntarily produce a GHG inventory of its 360 power plants in 1999–2000 and to establish a Carbon Fund to facilitate emissions reduction projects and trading (Kotov and Nikitina 2003, 18). The company prepared itself to benefit from the take-off of carbon markets and this preparation was made long before the Russian government was able to formulate a coherent national policy on the Kyoto Protocol. In 2003, UES joined the Global Greenhouse Gas Register of the World Economic Forum. The company's leaders had come to appreciate that the Protocol's flexible mechanisms could support their objective of attracting foreign investments to facilitate the modernization of electricity production. As a consequence of the inefficiencies inherited from years of centralized economic planning and outdated infrastructure, multiple opportunities were identified for cost-effective JI projects involving efficiency improvement, modernization, and increasing the share of natural gas and renewable sources in electricity generation (Muller 2004; Korppoo et al 2006).

Positive engagement in the climate regime was also important for the international image of the electricity utility, given its strategic interest in attracting foreign investment to finance its modernization and restructuring programme (Chubais 2001). European companies, as major potential investors, were particularly sensitive to climate issues as the EU prepared for the adoption of its Emission Trading Scheme. The EU's commitment to implement the flexible mechanisms and Japan's growing interest in JI further strengthened UES' interest in the Kyoto Protocol. The electricity monopoly thus became one of Russia's most vocal political supporters for the ratification of the treaty and, subsequently, of its implementation.

The natural gas monopoly Gazprom is another energy giant that lent political support to Russia's participation in the Kyoto Protocol. Gazprom is a joint stock company with more than 500,000 stockholders and the state holding a controlling share (50.002 per cent).[17] Gazprom is the largest producer of natural gas in the world. The company controls about 90 per cent of Russia's natural gas, operates the country's gas pipeline network and accounts for approximately 10 per cent of Russia's GHG emissions (Moe and Tangen 2000; O'Brian 2004). It is the single largest exporter in Russia, providing approximately 25 per cent of the federal tax revenues.[18] The EU is one of the company's most important international markets. According to Gazprom's website, it supplies about 'one third of Western Europe's total gas imports.'[19] Since natural gas is the least carbon intensive fossil fuel, it is estimated that the entry into force of the EU Emissions Trading System alone could boost European demand for natural gas by as much as 6 per cent in the period 2008–2012, since fuel switching is one of the cheaper, short-term options for industrial enterprises to meet their CO_2 emission caps (Grubb 2004, 6). Gazprom could further expect to benefit from emissions trading and JI by rationalizing its production and transmission to realize low cost emission reductions. According to an estimate by Moe and Tangen (2000, 84–85), leakages of methane from production, high pressure trunk pipelines and compressor stations account for more than half of the company's GHG emissions.

Gazprom, similar to UES, also had reputational incentives to support the climate regime as a mechanism to increase its attractiveness for foreign investment. Not coincidentally, one of the first pilot energy savings projects of Gazprom was implemented in cooperation with the German company Ruhrgas AG in 2005–2006 for the optimization of gas transportation networks, and reduction of fuel gas waste, consumption and emissions. Gazprom's annual environmental report highlighted the project not only for its emission reduction benefits, but also as 'an important driver for strengthening and improving production and commercial ties between the companies' (Gazprom 2006, 35).

The change in Gazprom and UES' interests with respect to climate cooperation created a powerful industrial lobby in support of the Kyoto Protocol. The rules of the environmental regime, the commitment of EU countries and Japan to implement the flexible mechanisms, and these companies' reputational concerns informed this change in strategy considerably *prior* to President Putin's decision to endorse the treaty. Although controlled by the state, these energy companies and their leadership also had a significant influence in the presidential administration,

---

[17] See <http://www.gazprom.com/eng/articles/article21713.shtml>, accessed 24 July 2008.

[18] US Energy Information Agency, <http://www.eia.doe.gov/emeu/cabs/Russia/pdf.pdf>, accessed 24 August 2008.

[19] <http://www.gazprom.ru/eng/articles/article20160.shtml>, accessed August 24, 2008.

illustrating the complex and close relations between business and government in Russia. The Head of the Management Board of UES, Anatoly Chubais and the former Chairman of the Board of Gazprom and current President of Russia, Dmitry Medvedev, were insiders in the administration. Medvedev was a particularly influential figure as Vladimir Putin's Chief of Staff between 2003 and 2005.

The support of UES and Gazprom for the Kyoto Protocol was representative of a broader and enlarging business coalition, including diverse actors such as smaller industrial enterprises, financial institutions, the National Carbon Sequestration Foundation, the National Carbon Union and the Union of Russian Industrialists and Entrepreneurs (which is Russia's largest industrial lobby) (Karas 2004; Murphy 2003). By early 2004 the interest and influence of economic actors had created a political base for the ratification that was absent in the early years following the adoption of the Kyoto Protocol.

Interplay of international and domestic politics in the Kyoto Protocol implementation

The Kyoto Protocol had an impact on interest group politics, bureaucratic realignment and economic epistemic consensus in Russia which were critical domestic factors also for the subsequent implementation of the treaty. Table 2 summarizes the main mechanisms of international influence and their effect on Russia's climate politics. Of these mechanisms, the expectation of positive distributional benefits was by far the most important for the realignment of domestic interests and bureaucracies in favour of the Protocol. However, capacity building and diffusion of economic expertise were instrumental in helping actors to assess the benefits of the flexible mechanisms, strengthen their institutional base to participate in these mechanisms and influence the policy discourse.

The analysis thus demonstrates that mechanisms of international influence, as identified in alternative, competing theoretical perspectives, often work in tandem with domestic processes. Indeed when socialization, distribution

Table 2. Impacts of the international climate regimes on Russian politics

Causal mechanisms	*Impact*
Scientific consensus	Limited: gate-keeping role of sceptics as presidential advisors
Norm diffusion	Limited: no direct access for NGOs to decision-making; weak public concern
Economic expertise	Change in issue-framing: policy discourse is focused on the economic benefits of the Kyoto Protocol
Technical assistance, training	Supported the positioning of the Ministry of Economic Development and Trade as the lead agency in climate policy; facilitated issue-framing—linking trade with Kyoto Protocol implementation
Distributional benefits associated with the flexible mechanisms and excess emission reduction allowance	Significant realignment of key industrial interests and coalitions in support of the Kyoto Protocol

incentives and capacity building complement each other, the impact of international institutions on domestic politics and state behaviour is likely to be the greatest. Furthermore, these mechanisms interact with domestic politics and produce consequences that are dynamic, and could not be fully anticipated by a unitary actor analysis. This is especially true in the field of climate policy, which links multiple regulatory instruments whose distributional implications depend on evolving markets, technology and institutions. From a unitary analysis perspective, for example, it would be difficult to understand why Russia questioned for so long the very economic benefits for which it bargained so hard in Kyoto. Russia's long wavering can only be understood if one takes into account its centralized decision-making structure and the initial gate-keeping role of climate sceptics in the presidential administration, coupled with low public interest and a lack of clarity on the feasibility and economic value of the flexible instruments.

A unitary actor analysis is similarly ill-suited to anticipate the turnaround of major energy companies into supporters of climate cooperation. These domestic changes unfolded in the course of several years between 1999 and 2004, ultimately influencing President Putin's decision to endorse the Kyoto Protocol via powerful insiders in the presidential administration and the active role played by the Ministry of Economic Development and Trade. As expected given Russia's 'superpresidential system', the Kyoto Protocol was ratified by the Federal Assembly in October 2004 within several months of the presidential endorsement.

The significance of domestic political changes is even more evident in the relatively swift adoption of procedures for the implementation of the Protocol after years of indecision on ratification. In early 2005, the government adopted a National Action Plan and created an Interagency Commission on the Kyoto Protocol, chaired by the Ministry of Economic Development and Trade. In August 2005, the Ministry started drafting the regulatory framework for the implementation of the flexible mechanisms. Despite some disagreements and turf battles among different agencies participating in the Commission, all implementing regulations were in place by June 2007. Russia announced its first 25 JI projects ready to be considered by the Climate Change Convention approval process.[20]

Due to their early technical preparation, commercial actors were ready to proceed with project implementation and pressed the government for timely regulatory action. UES had 30 JI projects ready for registration by 2006, and the company's Energy Carbon Fund had identified 103 potential projects with a primary focus on energy efficiency and renewable sources (Karppoo et al 2006). In March 2008, the UES Energy Carbon Fund put up for auction energy reduction units to the amount of 550,000 tons of CO_2, gained from the modernization of its Kirishskaya Thermal Power Plant. The highest bid was made by the UK-based Clean Planet Group for a price 'estimated at more than 4.5 million Euro'.[21]

[20] <http://unfccc.int/files/press/news_room/press_releases_and_advisories/application/pdf/20070601_english.pdf>, accessed 24 July 2008.

[21] <http://www.rao-ees.ru/en/news/pr_depart/latest_pr/show.cgi?030308rao.htm>, accessed 7 March 2008.

Gazprom has been similarly active in preparing to take advantage of global carbon transactions. In January 2007, it established a joint venture with Dresdner Bank of Germany to invest in various projects, which according to the *Financial Times* could generate as much as $19.4 billion in carbon credits (Harvey and Crooks 2007). Several months later, its London subsidiary, Gazprom Marketing and Trading, signed an agreement to purchase carbon credits from Propower do Brasil, a renewable energy plant, which could be resold on the European market (Kramer 2007).

The interest of large industrial and commercial actors to engage in GHG emissions reporting, reductions and trading has influenced profoundly the government's assessment of Russia's ability to implement and benefit from the Kyoto Protocol. Russia's 2006 National Communication, for example, anticipates approximately 10 per cent reduction in methane emissions between 2005 and 2012 just as a result of measures to be implemented by Gazprom (Russian Federation 2006, 64). The document further points to planned emission reductions by UES, by other industries and by the residential sector along with international projects as the main mechanisms to realize the country's climate policies and commitments (Russian Federation 2006).

Conclusion

Climate cooperation is at a critical juncture as countries begin talks on a successor agreement to the Kyoto Protocol. The preceding analysis of Russia's evolving politics in the context of the climate regime provides not only theoretical insights into the multiple mechanisms of regime influence, but also important policy implications. Russia's recent support for the climate regime can be maintained as long as positive distributional implications and domestic coalitions spanning economic interest groups and top government bureaucracies persist. The newly elected President Dmitry Medvedev is likely to maintain a pragmatic view seeking a linkage between technology investments, carbon markets and the stabilization of GHG emissions. As the salience of climate cooperation increases for G8 countries and the world, Russia will want to maintain a strong presence in this dynamic foreign policy arena.

In the upcoming rounds of negotiations, Russia could indeed find itself again in a balancing position between the EU's preferences for ambitious GHG emission reductions and the still uncertain, and likely cautious, position of the US. Like the US, Russia strongly supports emission caps for developing countries. It also insists on linking participation in the flexible mechanisms to the adoption of emission caps and compliance with the requirements of the Climate Change Convention. This reflects concerns about economic competitiveness with respect to large emerging economies, as well as what Russia sees as unfair competition and a crowding out of emission reduction projects.

The domestic politics analysis also implies that Russia would seek to maintain some cushion of 'hot air' in the next commitment period. Although the EU might be reluctant to grant 'hot air', this might be the only way to keep Russia engaged as well as to motivate developing countries to adopt binding emission ceilings through allowing headroom for growth and providing incentives to keep emissions lower than their targets, in order to sell the unused amounts on the carbon market. The Russian case suggests that there may be a silver lining in using 'hot air' as a medium-term instrument to create domestic support for climate cooperation in societies characterized by low environmental concern, and to spur interest in efficiency, technology innovation, as well as to decouple economic growth from emissions. Figure 2 reveals that Russia's impressive economic growth since 1999 significantly outpaced the rise of GHG emissions, which in 2005 were still 29 per cent lower than its Kyoto target. This was largely a result of economic restructuring and efficiency gains. If the country succeeds in maintaining the trend of decoupling economic and emissions growth, it could have a significant margin of excess emission reduction units at the end of the first commitment period, allowing it to gain from carbon trading and sustain domestic political support for the adoption of new and more ambitious targets. This outcome would mean that, despite the lenient terms of the Kyoto Protocol for Russia, the agreement has helped redirect the country towards a cleaner future.

For such a scenario to materialize and persist into the future, however, a strong linkage between foreign investment and emissions reductions has to be maintained, along with the leadership of industrialized countries including the US, and continued capacity building to ensure the transparency of carbon transactions and investments. It would be similarly important to broaden the base of domestic actors in Russia interested and able to participate in carbon markets and climate politics. This article reveals the limited role of international socialization in increasing broad public concern about climate change in Russia. However, the increased scientific attention placed on the ecological and human vulnerability of permafrost areas and other sectors can generate broader domestic concerns in countries like Russia. This is because such studies along with capacity-building programmes focused on adaptation can provide mechanisms for communicating climate risks more directly to specific sectors of society and increasing their capacity to meet those risks. The climate change regime must ultimately help societies, not just states, develop and deal with a global risk through markets, information and capacity. Nothing short of such a broad effort will succeed.

References

Agency for Direct Investments of the Russian Federation (2004) 'Proposal on procedures of registration and implementation of Joint Implementation projects in the Russian Federation', Presentation at the workshop on Joint Implementation, Moscow, 26–27 May

Anisimov, Oleg and Svetlana Reneva (2006) 'Permafrost and changing climate: the Russian perspective', *Ambio*, 35:4, 169–175

Avdeeva, Tatiana G (2005) 'Russia and the Kyoto Protocol: challenges ahead', *Review of European Community and International Environmental Law*, 14:3, 293–302

Botcheva, Liliana and Lisa L Martin (2001) 'Institutional effects on state behavior: convergence and divergence', *International Studies Quarterly*, 45:1, 1–26

Brewer, Thomas L (2007) 'Public opinion on climate change issues in the G8+5 countries', Working Paper, Georgetown University, Washington DC

Brown Weiss, Edith and Harold K Jacobson (1998) *Engaging countries: strengthening compliance with international environmental accords* (Cambridge, Massachusetts: MIT Press)

Checkel, Jeffrey (2001) 'Why comply? Social learning and European identity change', *International Organization*, 55:3, 553–588

Chubais, Anatoly (2001) 'Russian power sector reform', Presentation by Anatoly Chubais, Chief Executive Officer, RAO UES Russia, Moscow, 19 June

Churchill, Winston (1939) 'The Russian enigma', Radio broadcast on 1 October 1939, <www.churchill-society-london.org.uk/RusnEnig.html>, accessed June 2006

Colton, Timothy J (2007) 'The enigmatic Mr Putin', *Harvard Magazine*, 109:5, 40–43

Easter, Gerald M (1997) 'Preference for presidentialism: postcommunist regime change in Russia and the NIS', *World Politics*, 49:2, 184–211

Gazprom (2006) 'Environmental Report 2006', <http://www.gazprom.com/documents/Ecology_Eng.pdf>, accessed 21 July 2008

Gourevitch, Peter (1978) 'The second image reversed: the international sources of domestic politics', *International Organization*, 32:4, 881–912

Grubb, Michael (2004) 'Russian energy and CO_2 emission prospects: evidence from domestic analyses and international comparisons', Briefing Note, Royal Institute of International Affairs, London

Haas, Peter (1989) 'Do regimes matter? Epistemic communities and Mediterranean pollution control', *International Organization*, 43:3, 377–403

Harvey, Fiona and Ed Crooks (2007) 'Gazprom launches carbon trading venture', *Financial Times*, 15 January

Henry, Laura A and Lisa McIntosh Sundstrom (2007) 'Russia and the Kyoto Protocol: seeking an alignment of interests and image', *Global Environmental Politics*, 7:4, 47–69

Karas, Jacqueline (2004) 'Russia and the Kyoto Protocol: political challenges', Briefing Note, Royal Institute of International Affairs, London

Keck, Margaret and Kathryn Sikkink (1998) *Activists beyond borders: advocacy networks in international politics* (Ithaca: Cornell University Press)

Korppoo, Anna, Jaqueline Karas and Michael Grubb (eds) (2006) *Russia and the Kyoto Protocol: opportunities and challenges* (London: Chatham House)

Kotov, Vladimir (2002) 'Policy in transition: new framework for Russia's climate policy', Working Paper 58, Fondazione Eni Enrico Mattei, Milan

Kotov, Vladimir and Elena Nikitina (2003) 'National Framework for GHG Emission Trading in Russia', Paper prepared for the OECD Global Forum on Sustainable Development: Emissions Trading held in Paris, CCNM/GF/SD/ENV(2003)8/FINAL, 17–18 March

Kramer, Andrew E (2007) 'Gazprom shifts its weight to carbon trading', *The New York Times*, 24 April

Miles, Edward L, Arild Underdal, Steinar Andresen, Jorgen Wettestad, Jon Birger Skjaerseth and Elaine Carlin (eds) (2002) *Environmental regime effectiveness: confronting theory with evidence* (Cambridge, Massachusetts: MIT Press)

Milner, Helen V (1988) *Resisting protectionism: global industries and the politics of international trade* (Princeton, New Jersey: Princeton University Press)

Moe, Arild and Kristian Tangen (2000) *The Kyoto mechanisms and Russian climate politics* (London, United Kingdom: Chatham House)

Muller, Benito (2004) 'The Kyoto Protocol: Russian opportunities', Briefing Note, Royal Institute of International Affairs, London

Murphy, Kim (2003) 'Polluters join call for Russia to ratify Kyoto pact', *Los Angeles Times*, 5 September

Myers, Steven Lee and Andrew Revkin (2003) 'Russia to reject pact on climate, Putin aide says', *The New York Times*, 3 December

Nikitina, Elena (2001) 'Russia: climate policy formulation and implementation during the 1990s', *Climate Policy*, 1:3, 289–308

O'Brian, John (2004) 'Russia and Kyoto: match made in heaven?', Presentation to COP10 Side Event 'Unlocking Russia's JI Potential' at Buenos Aires, Argentina, 15 December, <http://www.ieta.org/ieta/www/pages/getfile.php?docID = 703>, accessed 19 July 2008

Putnam, Robert (1988) 'Diplomacy and domestic politics: the logic of two-level games', *International Organization*, 42:3, 427–460

Rogowski, Ronald (1990) *Commerce and coalitions: how trade affects domestic political alignments* (Princeton, New Jersey: Princeton University Press)

Russian Federation (2002) ''Третье национальное сообщение [Third national communication to the UNFCCC]', <http://unfccc.int/resource/docs/natc/rusncr3.pdf>, accessed 19 July 2008

Russian Federation (2006) ''Четвертое национальное сообщение [Fourth national communication to the UNFCCC]', <http://unfccc.int/resource/docs/natc/rusnc4r_rev.pdf>, accessed 19 July 2008
Schreurs, Miranda A and Elizabeth Economy (eds) (1997) *The internationalization of environmental protection* (Cambridge: Cambridge University Press)
Shevtsova, Lilia (2000) 'The problem of executive power in Russian', *Journal of Democracy*, 11:1, 32–39
Simmons, Beth A and Lisa L Martin (2002) 'International organizations and institutions' in Walter Carlsnaes, Thomas Risse and Beth A Simmons (eds) *Handbook of international relations* (London: Sage Publications), 192–211
Tallberg, Jonas (2002) 'Paths to compliance: enforcement, management, and the European Union', *International Organization*, 56:3, 609–643
Yakovlev, Andrei (2006) 'The evolution of business-state interaction in Russia: from state capture to business capture?', *Europe-Asia Studies*, 58:7, 1033–1056
World Bank (1999) *Study on Russian national strategy of greenhouse gas emissions reduction* (Washington, DC: World Bank)
World Bank (2007) *World development indicators online 2007* (Washington, DC: World Bank)

The construction of China's climate politics: transnational NGOs and the spiral model of international relations

Miriam Schroeder

Introduction

In the inglorious ranking list of the world's top greenhouse gas (GHG) emitters, China is catching up rapidly. According to recent publications, China has overtaken the United States' (US) CO_2 emissions by 14 per cent in 2007 and has become the world's largest CO_2 emitter, making up 24 per cent of the world's total CO_2 emissions (Netherlands Environmental Assessment Agency 2008). As China is becoming one of the crucial factors for the advancement of the international regime on climate change, it is challenged by both state and nonstate actors to curb its growing GHG emissions. Although China—in concert with other developing countries—still adheres to the principle of 'common but differentiated responsibilities', one can recognize substantial transformations in China's climate change politics at both the international and national level. China ratified the Kyoto Protocol in 2002, and the Chinese government is increasingly designing and implementing national-level policies with positive (side-)effects for climate protection. These measures to curb GHG emissions are, at least rhetorically, justified by the objective of climate protection, although many authors warn of a growing gap between ambitious rhetoric and ineffective policy implementation on the ground (Economy 2006, 2007; Turner and Linden 2007).

This article takes a constructivist approach and argues that the transformations in Chinese climate change politics can be explained by an internalization of the norm of climate protection by the Chinese government. The research question guiding this analysis is: *how has the norm of climate protection become internalized in Chinese politics?*

The focus of the analysis is on the role of Chinese and international nongovernmental organizations (INGOs) as norm advocates, the impact of their transnational cooperation on norm transfer and ultimate transformation of Chinese climate politics. Several structural causes, most prominently the rise of the market economy, decentralization and a 'greening' of the Chinese state (Yang 2005; Ho and Vermeer 2006) have provided a space for the emergence of environmental NGOs. Part of their success lays in the cooperation with international NGOs and scientific communities that also work for environmental protection. Although previous articles have already provided accounts of the patterns of cooperation which can be observed between INGOs and environmental NGOs (for example, Morton 2007; Yang 2005), this policy-field specific analysis of transnational NGO impacts on Chinese climate politics should provide theoretical lessons for constructivist explanatory approaches as well as practical considerations on effective NGO cooperation for climate protection.

In the first part of the article the constructivist approach of norm internalization is briefly introduced. The so called 'spiral model'—developed by Risse et al (1999) as a causal explanation for the transnational cooperation between NGOs—has proven valuable to understanding the human rights field. The model will be described and adopted for its application to the field of climate politics; thus testing the model to check whether its validity can be extended towards other fields of politics in which norms also impact political decisions. The second part of this article is empirical; in it the revised spiral model is applied to process trace the impact of international and Chinese NGOs on Chinese climate politics. As a complement to the longitudinal analysis, the article focuses on the present modes of interaction between INGOs and Chinese NGOs. The article concludes with some theoretical findings from the application of the spiral model to climate politics and draws lessons-learnt for the role of environmental NGOs in countries with a restricted civil society, especially in regard to their possibilities to link up with international civil society.

Taking a constructivist approach in explaining change in national climate politics: norm internalization

The assumption that a change in Chinese climate politics has been induced by an internalization of the international norm of climate protection marks this article as part of a constructivist approach to international relations. Typical for the constructivist approach is the assumption that norms are defining the interests of actors, thus providing the starting point for actions. The causal priority of norms for the explanation of interests is derived from the constructivist image of an actor and her impetus for acting. In contrast to rationalist schools of international relations, the constructivist approach conceptualizes a person as a *homo sociologicus*, an actor that is guided in his or her behaviour by the logic of appropriateness and a bounded rationality. Behaviour according to the logic of appropriateness means acting according to one's own perceived identity and the expectations of one's norm community (March and Olsen 1989, 160–162).

In all cases of norm internalization, but especially in cases where international and national norms collide, norm entrepreneurs play an important role as actors transmitting and enforcing the international norms (Finnemore and Sikkink 1998, 896). Every entity of a norm community can act as a norm entrepreneur, but within the field of environmental politics, international organizations, epistemic communities[1] and nongovernmental organizations are the most common. All of these actors can be part of a transnational advocacy network, defined as 'those actors working internationally on an issue, who are bound together by shared values, a common discourse, and a dense exchange of information and services ...' (Risse et al 1999, 18; similar approaches in Risse-Kappen 1995; Keck and Sikkink 1998, 2). If domestic advocators of a norm cooperate with actors of the international norm community, the likelihood increases that this transnational advocacy network is able to make the norm-violating state succumb to the international norm (see Figure 1).

It is important to note the different strategies actors of the transnational advocacy network employ to transfer and enforce a norm. The spiral model mainly envisages norm entrepreneurs as using a mixture of pressure and shaming tactics, but other norm advocacy strategies such as information politics, symbolic politics, leverage politics and accountability politics are also an option (Keck and Sikkink 1998, 6). Which strategy is chosen and proves most successful depends on the framing of the topic under discussion, the campaigning history of the norm entrepreneur and the domestic institutional conditions of the country under consideration. For example, climate change framed either as an environmental or as an energy issue appeals to different audiences and calls for different types of advocacy strategies.

Introduction to the spiral model of transnational norm diffusion

Starting from the research interest to explain the variance of changes in human rights politics on the national state level, the spiral model assumes a five-phase pathway of how norm internalization works in practice. These five phases are in theory sequential to each other, but may overlap in practice.

- *Phase One—repression and activation of network.* Facing a strong violation of a behavioural norm, a transnational network of actors forms for advocacy. Once the transnational network of norm entrepreneurs is activated, pressure tactics are used to force the repressive state into dialogue by imposing material pressure (for example in the form of trade sanctions) and/or by imposing social pressure (for example through shaming-tactics). It is assumed that the target state will respond with instrumental adaptation.

  Hypothesis for the transition to Phase Two: Only if and when the transnational advocacy network succeeds in gathering sufficient information on the norm violation by the target state, can it put the norm-violating state on the international agenda, thus moving the situation to Phase Two.

[1] Epistemic communities are defined according to Haas (1992, 3) as 'a network of professionals with recognized expertise and competence in a particular domain and an authorative claim to policy-relevant knowledge within that domain or issue-area'.

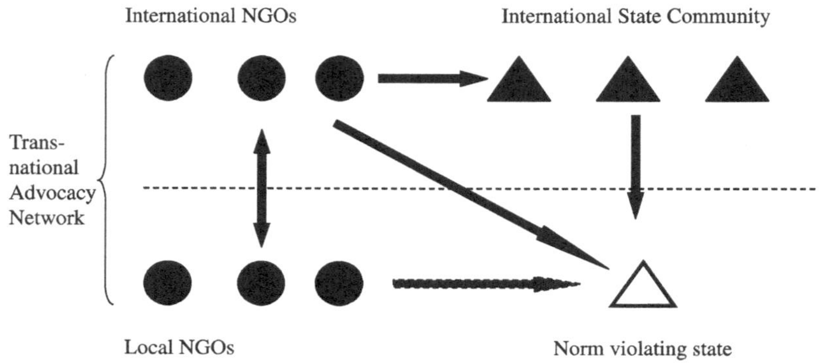

Figure 1. The transnational advocacy network

- *Phase Two—denial.* Domestic norm entrepreneurs enter the scene and lobby together with their international counterparts for the norm under consideration. In this phase, denial of the validity of the norm is regarded as progress since it signals the target state beginning to be involved in a dialogue. Entering a dialogue might either be motivated by social or material vulnerability.

  Hypothesis for the transition to Phase Three: If the strength of the transnational advocacy network and the vulnerability of the target state facing international pressure are large enough, the target state makes first tactical concessions.

- *Phase Three—tactical concessions.* This phase is characterized by rhetoric and practical norm acknowledgements by the target state. Due to these—in most cases tactical—concessions, a window of opportunity is opened for the transnational advocacy network to increase its pressure. Once trapped in its rhetorical norm acknowledgements, the state has to continue on the involuntarily chosen path of norm internalization.

  Hypothesis for the transition to Phase Four: If faced with a fully mobilized domestic opposition, linked up with transnational networks for which the norm under consideration has reached consensual status, norm-violating governments no longer have many choices and usually become self-entrapped into argumentative behaviour in favour of the norm.

- *Phase Four—prescriptive status.* This phase describes a situation in which the target state justifies its own behaviour in accordance with the norm, so that its validity is no longer disputed, even if norm violations might still happen in some instances. One difficulty in identifying this phase is the problem of how to measure norm change, especially how to distinguish norm internalization due to real persuasion from rhetorical and instrumental adaptation. Risse et al (1999) take a pragmatic view by arguing that true beliefs of actors are not crucial as long as the actors are consistent in their verbal utterances and their words and deeds ultimately match.

  Hypothesis for the transition to Phase Five: If the pressure from the domestic civil society and the transnational advocacy network remain constant, the target state holds on to its concessions leading to long-term norm-consistent behaviour.

- *Phase Five—rule-consistent behaviour.* If the pressure from international and national norm entrepreneurs is continuing, the final state of the norm internalization process—in which international norms are fully institutionalized domestically and norm compliance becomes a habitual practice of actors and is enforced by the rule of law—is reached.

The authors of the spiral model provide several intervening variables that might influence the norm internalization process. For the present analysis of climate change politics, only one of them—world time—is relevant: it describes the intervening impact of the international cascade of a norm upon the national level. If an international norm is followed by an increasing number of members of the international community, the ease and likeability increases of it becoming internalized by any other state.

Applying the spiral model to environmental politics

The spiral model has been developed as 'a theory of the stages and mechanisms through which international norms can lead to changes in behaviour' (Risse et al 1999, 2) in order to explain the influence of international norms on national human rights politics. Because we can also identify norms in the field of environmental politics, it seems plausible to transfer the model from human rights to the environmental politics field. This should also be what the developers intended, since they state that the model could be used for any sociopolitical processes of change involving norms independent from its geographical region and national setting (238).

For a long time conflicts in the field of environmental politics have been primarily discussed as interests and distribution conflicts about absolute and relative goods with 'tragedies of the commons'[2] and 'collective action problems'[3] being the keywords of the debate. Like in other fields of politics the shortcomings of such rationalist approaches lay in their inability to explain a change in behaviour due to the change of norms. Rationalist approaches are also facing the difficulty of explaining the creation and change of preferences in the field of environmental politics. An increasing number of authors are now looking at the impact of international and national norms as reasons behind actions taken in environmental politics (for example Keck and Sikkink 1998; Stokke 1998). An explanatory approach focused on the internalization of norms seems fruitful, especially for the explanation of a long-term change of national and international behaviour in the field of environmental politics.

If one compares different cascades of norms, the norm of climate protection has been one of the most prominent in the last decades. Climate protection first

---

[2] 'Tragedies of the common' describes a situation in which damage caused by the action of one actor to the whole community surpasses the utility of the action for the actor causing it (Hardin 1968).

[3] 'Collective action problems' exist when a cooperative solution to a problem is beneficial to all actors but would open the possibility for one or several actors to free-ride on the actions (and benefits) carried out by only a few actors.

became widely recognized at the first World Climate Conference in 1979, from this date onwards a very dynamic international cascade of the norm of climate protection can be observed. The UN Framework Convention on Climate Change (UNFCCC) has become the universal agreement of 154 countries. The norm to protect the climate is spelled out as 'achiev[ing] stabilization of greenhouse gas concentrations in the atmosphere at a low enough level to prevent dangerous anthropogenic interference with the climate system' (article 2), and is comparable in its scope to the Universal Declaration of Human Rights. Both norms are comparable in their development—human rights have reached earlier consensus but still face violations, just like the norm of climate protection. Precise measurement is difficult for all instances of norm change, because rhetoric and behaviour do not always match. For the following case study, norm change is measured in two steps: first, through changes in rhetoric by leading Chinese governmental officials in national and international discourses, and second, by changes in national and international policies relevant to climate change.

Case study: explaining national change in Chinese climate politics

The People's Republic of China (PRC) is a prominent example for national politics that reflected rivalry between national and international norms for decades. It thereby also fulfils the two criteria, provided by Risse et al (1999, 273), required for the applicability of the spiral model in other political fields. First, a state needs to adhere to a particular set of public norms that have become embedded in its laws, institutions and policies; and second, these state norms need to be challenged by a contradictory set of international norms promoted by an emerging transnational advocacy network. As outlined above, the contrasting international norm of environmental protection can be regarded as an international consensus since 1972, and the norm of climate protection has occurred on the political agenda since 1979. Therefore it makes sense to begin the analysis of Chinese climate politics—embedded in the overarching field of energy and environmental politics—from that date in order to trace changes in behaviour and impacts of the transnational advocacy network. The following paragraphs provide a brief summary of the different phases of norm internalization in China arranged according to the spiral model.

Process tracing the impact of the transnational advocacy network on Chinese climate-change politics

Phase of 'no debate' (1949–1986)

At the end of the 1970s, when the topic of climate change was first raised on the international agenda, China had been in a state of struggle against nature since almost three decades, advocating a development strategy of resource-intensive growth. The Maoist negligence of nature can be partly explained by the utopian urgency during the Great Leap Forward which rapidly advanced industrialization and agricultural production, and partly by the communist credo that environmental problems are the creations of capitalist countries and are therefore, by definition, nonexistent in China (Ross 1988, 8; Duan et al 2000, 7). Driven by the urge for economic growth, energy prices were heavily subsidized (Hatch 2001, 9) leading to a doubling of energy demand in comparison to economic growth in the period 1949–1979 (Levine 1999, 6). With Deng Xiaoping's opening politics after 1978, the emphasis continued to be on resource-intensive economic growth, but its ideological targets shifted towards a modest prosperity for the whole society. One can therefore conclude that the national norm of how to approach natural and energy resources stood in stark contrast to the emerging international norm of climate protection, which calls for a sustainable use of natural and, especially, energy resources.

On the international stage, China did not participate in the first World Climate Conference in 1979. However, China did send a delegation to the first UN Conference on the Human Environment in 1972, where Chinese delegates voiced their opinion that communist countries do not face any environmental problems, while they blamed imperialist countries for their resource exploitation. The delegates reinforced their demand that the international norm of 'sovereignty with no interference in internal affairs' should be applied to environmental matters as well (Ross 1988, 137).

China's official stand in environmental negotiations changed slightly in 1973 with Prime Minister Zhou Enlai opening China's first National Conference on Environmental Protection and the establishment of the Environmental Protection Leadership Group, which would later become the National Environmental Protection Agency (NEPA)—the forerunner of the Ministry of Environmental Protection. According to Elizabeth Economy (1997, 22) the Prime Minister had been impressed by the report of the UN Conference on the Human Environment. At the second National Environmental Conference, members of NEPA already began to call for a drastic emission reduction in soot and sulphur dioxide to avoid further air pollution (Ross 1988, 143).

Phase of denial (1987–1990)

In 1987 the first sign of a Chinese engagement in the international debate on climate change can be detected; on 19 August an agreement between the US Department of Energy and the Chinese Academy of Science was signed for a joined research program about the impacts of carbon dioxide on climate change. As a consequence of this agreement, the United States of America (USA)-PRC Committee for the Joint Study of Global Change was founded and an exchange of scientists began to take place. In contrast to the causal mechanisms set forth by the spiral model, not advocacy organizations but scientists were the first to put the topic of climate change on the Chinese agenda. This can be explained partly by the issue itself, partly by the lacking strength of the transnational advocacy network in the 1970s.

In line with the bilateral scientific exchange, a Chinese National Climate Committee was founded in 1987 to coordinate Chinese research and examine the effects of a possible climate change for China. China even took part in the first meeting of the International Panel on Climate Change, but still denied all responsibility and need for action (Economy 1997, 29). Shortly before the Second World Climate Conference in 1990, an internal debate about China's official position took place. There were differences among the conservative Ministry of Foreign Affairs, the Planning Commission and the Energy Ministry, and the rather progressive representatives of the State Science and Technology Commission and

NEPA. Although the arguments on the conservative side reflected the traditional line, voices from the progressive fraction—for the first time—raised the factor of a possible image boost by a constructive Chinese position, which the country was in need of after its heavy loss of reputation due to the 1989 Tiananmen incident.[4] The debate was further intensified by the results of the report of Working Group II of the International Panel on Climate Change (IPCC), which placed China among the most vulnerable countries to climate change, but also listed possible positive effects for agriculture yields in Northern China. The debate was ultimately decided by the intervention of Deng Xiaoping, Leader of the Communist Party of China, and Li Peng, Premier of the State Council, who favoured the conservative position that regarded energy security as China's utmost priority (Economy 1997, 28). The phase of denial ended with China's signing up to the Montreal Protocol for the Protection of the Ozone Layer in 1990. This participation can be seen as the first cooperative move of the Chinese government in its contribution to the international endeavour of climate protection. On the national level, the eighth five-year plan (1991–1995) included—for the first time—the term of climate change when it called for a research project on the 'changes in the life-supporting environment in the next 20 to 50 years ...' (CSCPRC et al 1992, 16). This transition from the denial phase to the phase of tactical concessions cannot be explained by the actions of the transnational advocacy network (which at that time did not work in or on China). Instead, information shared by scientists, for example the results of the IPCC Working Group II about possible negative outcomes of a climate change for China, as well as China's urge for an improved image on the international stage can be detected as the main driving factors behind China's changing position on climate change.

Phase of tactical concessions (1991–2001)

On the international level, China began to switch to an active role in the climate negotiations, hosting the Beijing Ministerial Conference on Environment and Development in developing countries on 18–19 June 1991. The focus of this conference was indeed a debate on norms, '[to] discuss in depth the challenges faced by the international community in establishing norms of cooperation for the enhancement of environment and development, particularly the implications for the developing countries ...' (*Beijing Review* 1991, 9). However, the Declaration of the Beijing Conference still reflects the traditional viewpoint of developing countries—that climate change is the sole responsibility of developed countries and developing countries should not face any demands for action on climate protection. The consolidation of the developing countries' positions can be regarded as a prime purpose of the conference, with developing countries speaking on behalf of the Group of 77 and China, with China no longer being in an isolationist position (Economy 1997, 34). There was, however, a switch from an opportunistic position towards a more pragmatic one of the middle ground, signalling some willingness to cooperate. China signed the UN Framework Convention on Climate Change as the tenth country and ratified the convention on 2 November 1992. However, the fact that the State Meteorological Agency was

[4] This assessment is shared by Johnston (1998, 558) and Hatch (2001, 22).

assigned as the implementing agency reflects the Chinese handling of the climate change issue as merely a natural science phenomenon and not a political problem with severe social and economic consequences.

The UN Conference on Environment and Development in 1992 brought another stimulus to the debate about the appropriate path for China's development. With arguments ranging from deep ecology to 'no interference in sovereign matters', the debate within academia tended towards a more sustainable development strategy.[5] Even if most articles opted for a sustainable development strategy out of reasons of national interest—for example advocating that a decoupling of economic growth from energy demand would lead to improved energy security—some scientists also stressed the responsibility of China, as a growing superpower, to do its part for the solution of global environmental problems. This shift of opinion within the Chinese academia can be attributed to their increasing cooperation with international epistemic communities working on the climate issue. Many exchange programmes and events took place in the 1990s, often sponsored and organized by scientific institutes and international NGOs. For example, the World Wide Fund for Nature (WWF) organized an international seminar on global warming in China (Johnston 1998, 571); the Climate Institute coordinated a meeting between Chinese and international scientists (CSCPRC et al 1992, 190); and representatives of the Chinese State Science Commission cooperated with the UN Centre for Science and Technology Development (Johnston 1998, 571). Information and arguments provided by the international scientific community have helped to frame the climate issue in a way conducive to Chinese scientists and decision-makers, some of whom had become increasingly aware of the deteriorating state of the Chinese environment and its consequences for the overall well-being of the country. The intersection between the Chinese epistemic communities and political decision-makers is hard to follow from the outside, but China experts agree that the flow of information and arguments between the two communities can be considered high and influential on policy-making (Heggelund 2007, 169).

Concerning the participation of civil society in the Chinese environmental and climate politics, two trends are identifiable for the early 1990s. First, the Chinese government launched educational campaigns to strengthen the public's environmental awareness and knowledge; worldwide environmental events such as Earth Day and the World Environmental Day were introduced. The media coverage of official environmental activities and conferences was also enhanced by press campaigns such as 'Environmental Campaigns to the New Century' (Wen 1998, 39). Second, the first domestic environmental NGOs appeared on the Chinese scene. The 'Friends of Nature' (FON) were founded as China's first nature conservation organization in 1993. In the late 1990s several other civil-society-based organizations emerged in China, mainly focusing on public education campaigns for nature conservation. At present, China does not have an NGO solely dedicated to climate protection, but many NGOs have climate-related campaigns.

---

[5] Different positions on the question of whether to substitute the traditional Chinese development model with a model of sustainable development can be found in Xiao (1995), Cai (1997a, 1997b), Zhao (1998) and Liu (1995).

The phase of tactical concessions can be said to end with the ratification of the Kyoto Protocol on 30 August 2002. Although China as a developing country has no obligations for GHG emission reductions, the signing and ratification of the Kyoto Protocol is one indicator that the norm of climate protection has become inscribed in Chinese politics.

Phase of prescriptive status (2002–present)

The late 1990s also saw the uptake of emission reductions into the national legislation. Examples are the 1995 revision of the regulation on avoidance and control of air pollution, in the ninth five-year plan, which called for a decoupling of economic growth and environmental degradation and the tenth five-year plan, which included the target to reduce CO_2 emissions by 10 per cent in 2005 compared to 2000 levels (IEA 2007, 310). Although policies like these show a shift in Chinese policy-making towards emission reductions, they have not been linked rhetorically to climate protection but to domestic problems of air pollution and energy shortages. In line with industrial modernization, energy intensity had dropped during most of the 1980s and 1990s, but started to rise again between 2002 and 2004 (261). An official report estimated environmental degradation to have caused a 3 per cent loss of GDP in 2004 (NBS 2006). Putting a stop to environmental degradation and increasing energy efficiency has thus climbed to the top position of the political agenda, culminating in the target of the 11th five-year plan to reduce energy intensity by 20 per cent until 2015, and to increase the share of renewable energies in electricity generation from 8 per cent in 2005 to 10 per cent in 2010, and 16 per cent in 2020 (Zhang 2005). In addition, the law for the promotion of renewable energies came into force on 1 January 2006, and governmental institutions with relevance for climate governance, for example the former National Environmental Protection Agency were upgraded into the State Environmental Protection Agency, which now enjoys ministerial status.

On the international level, China has become the leading country for projects of the Clean Development Mechanism (CDM) of the Kyoto Protocol, has issued its National Climate Change Programme and set up the National Leading Group on Climate Change with Premier Wen Jiabao as its director in June 2007. The Chinese delegation engaged constructively in negotiations on the post-2012 Kyoto regime during the last Conference of the Parties (COP)/MOP[6] 13 meeting in Bali and the government welcomes the Bali Roadmap (Foreign Ministry of the PRC 2007). In addition to the Kyoto Protocol, China has also joined climate-related initiatives such as the Asian-Pacific-Partnership on Clean Development and Climate in January 2006 and the Major Economies Meeting on Energy Security and Climate Change in September 2007. In contrast to the assumptions of the spiral model, it had been government representatives from Organisation for Economic Development and Cooperation (OECD) countries and not the transnational advocacy network who from the mid-2000s started to press the Chinese government to accept further obligations under the Kyoto protocol. This mainly took the form of

[6] The COP is comprised of countries that have ratified or acceded to the UNFCCC and is the supreme governing body of the UNFCCC. The COP/MOP is comprised of Parties to the UNFCCC who are also Parties to the Kyoto Protocol.

addressing the topic of climate change during state visits, for example EU Commission President Barroso, Germany's Chancellor Merkel, and France's President Sarkozy all addressed climate change during their state visits to China in 2007. Allegedly, this pressuring strategy paid off, as one interviewee explained,

> Every time foreign leaders come to China, they want to discuss climate change. Every time Hu Jintao goes abroad, foreign leaders want to discuss climate change. It is really getting on his nerves. The Chinese delegation submitted this very constructive proposal to the Bali COP/MOP, because the leader had said 'you must do something to reduce the pressure on my shoulders'.[7]

Even if the trend in Chinese climate politics goes towards internalization of the norm of climate protection, the phase of norm-consistent behaviour has not yet been reached as several indicators, such as mechanisms for citizens to complain about norm-inconsistent behaviour, are established but not used (*People's Daily Online* 2006) and the reliance on laws to deliver environmental governance is still in an initial stage. Although there has not been much change in the official Chinese position in international negotiations for the Kyoto Protocol—China, in line with other developing countries, is not willing to take up any legally binding GHG emission reduction commitments—China is increasing its climate protection measures on the national level, although these are embedded in the discourse on energy security and local air pollution and not in that of global climate protection.

Focus on the role of transnational NGO cooperation

This section complements the previous historical process tracing of Chinese climate politics with a more in-depth analysis of how INGOs and Chinese NGOs can presently influence climate-related policy-making in China. The first part briefly outlines the situation of Chinese environmental NGOs and their positioning in climate politics; the second provides an analysis of NGO impacts on Chinese policy-making, which is based on 12 interviews with four international and eight Chinese NGOs conducted in early 2008.

The Tsinghua NGO Research Centre estimates the number of unregistered NGOs in China to be between 1.6–3 million (Brie and Pietzcker 2004, 7; Kuhn 2006). The border line between an NGO and government-organized NGOs (GONGOs) is vague because even an NGO that is financially independent from the government requires its support for its daily operations, creating almost a 'symbiotic situation' (Ho 2007, 198), in which NGO leaders and politicians are mutually dependent on each other. The Chinese government considers the actions of its domestic NGOs as acceptable as long as these are in line, or even supporting, official policies, such as a public education campaign on energy saving behaviour. Although the rules for the establishment and the registration for nongovernmental organizations originating from 1989 were revised twice, in 1996 and 1998, these revisions did not necessarily ease the launching of NGOs. Due to the difficult registration procedures, many Chinese NGOs choose to register as businesses or not to register at all. Consequently,

---

[7] Confidential interview with representative of International NGO E, 16 January 2008, Beijing.

Yang (2005, 50) differentiates between seven different types of environmental NGOs in China: registered NGOs, nonprofit enterprises (which operate as nonprofit organizations but are registered as business enterprises), unregistered voluntary groups, web-based groups, student environmental associations, university research centres and government-organized NGOs (GONGOs).

Environmental NGOs in China started as nature conservation organizations and watchdogs for local air and water pollution. The All China Environmental Foundation (ACEF), a GONGO working as a network for Chinese environmental NGOs, states the current number of environmental NGOs to be 2,768 (ACEF 2006). Out of these, interview partners were selected by two criteria: (a) the visibility of their climate-related activities on websites and conferences, based on the assumption that NGO visibility is one indicator for its soft power of influencing policy-makers, and (b) their recommendation by other NGOs based on the assumption that the circle of NGOs active in climate-related matters is small and mainly based in Beijing, as their target for influence is the central government (see Figure 2, which is the result of a mapping exercise conducted with the interview partners).

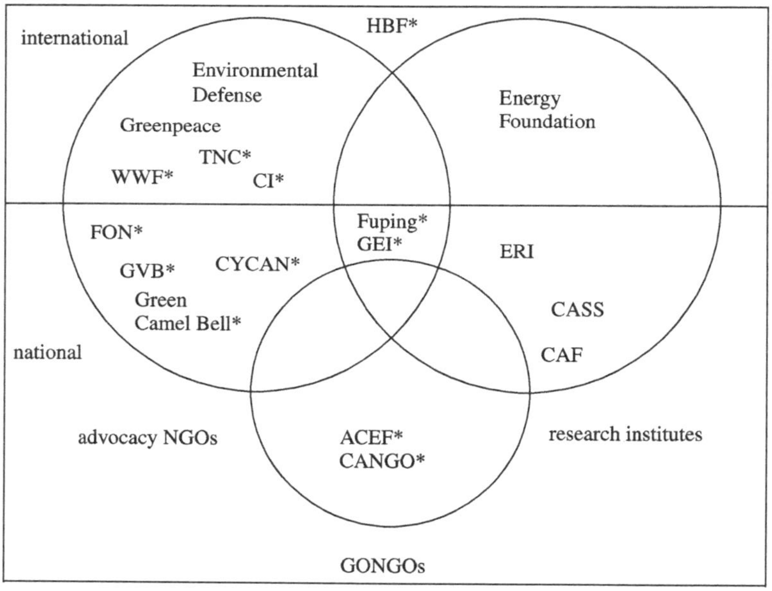

Figure 2. Mapping of NGOs and research institutes in China with climate-related activities
Source: Interviews 2007/2008. *Interviewed organizations: Environmental Defense Fund (www.edf.org); Greenpeace (www.greenpeace.org/china); WWF = World Wide Fund for Nature (www.wwfchina.org); TNC = The Nature Conservancy (www.nature.org/china); CI = Conservation International (www.conservation.org); FON = Friends of Nature (www.fon.org.cn); GVB = Global Village Beijing (www.gvbchina.org); CYCAN = China Youth Climate Action Network (www.cycan.org); Green Camel Bell (www.gcb.ngo.cn); Fuping Institute (www.fdi.ngo.cn); GEI = Global Environmental Institute (www.geichina.org); Energy Foundation (www.ef.org); ERI = Energy Research Institute (www.eri.org.cn); CASS = Chinese Academy of Social Science (bic.cass.cn); CAF = Chinese Academy of Forestry (www.forestry.ac.cn); ACEF = All China Environmental Federation (www.acef.com.cn); CANGO = China Association for NGO Cooperation (www.cango.org); HBF = Heinrich Boell Foundation China (www.boell-china.org)

The quantitative parts of the interviews are represented in the graphs (Figures 3–5), displaying absolute numbers of statements given in one answer category. In order to trace developments in transnational NGO cooperation over time, interviewees were asked to compare their present situation in 2008 with the situation five years ago in 2003. However, since some of the organizations interviewed were not yet established in 2003, data on this comparison remains sketchy. Despite the small number of interviews conducted, the interview statements give a first impression of what NGOs with climate-related activities perceive to be successful campaigns and replicable pathways of transnational cooperation.

The process tracing has revealed a change in the official Chinese position on climate change, which culminated in the ratification of the Kyoto Protocol in 2002 and in the energy saving and emission reduction target of the 11th five-year plan in 2005. This shift in governmental approach towards climate politics also made it easier for Chinese NGOs to address the issue openly: in the beginning of the millennium, NGOs framed their activities as energy saving campaigns, because the issue of climate change was still considered to be a sensitive political issue by the government, but since the mid-2000s several NGOs re-framed some of their activities as climate protection campaigns.[8]

Concerning the strategic choices for climate-related activities, China-based NGOs are actively using the media, publicizing own research material and conducting education campaigns to raise public awareness on climate change (compare Figure 3). Even if the framing of activities in the context of climate change has become politically feasible, most NGOs still prefer to frame their activities in the context of energy saving, because they feel they can more easily transfer this message to the public.[9] One of the best examples for the coordinated work of Chinese NGOs in the field of raising consumers' awareness of climate change and influencing their individual behaviour is the '26 Degree Campaign'. China usually experiences a peak in electricity consumption in the summer months during which air conditioners run at full power to cool down hotels, offices and private homes. The '26 Degree Campaign' was initiated in the summer of 2004 by six Chinese NGOs asking public and private enterprises as well as individuals to set their air conditioners to a minimum of 26 degrees. The achievements of the '26 Degree Campaign' has been a saving of 350,000 to 550,000 tons of CO_2 and the launch of 'The announcement about strictly implementing the standard of temperature control in air conditioners in public buildings' by the State Council in 2007, which was celebrated by the NGOs as partly being a result of their campaign.[10]

Cooperation among Chinese NGOs and international NGOs (INGOs) began in the late 1990s. The simplest form of transnational NGO cooperation is the funding of Chinese NGO activities by like-minded INGOs. This way, INGOs can have some decision-making power on local campaigns, but they do not need to go through the hassle to set up their own branches or found NGOs in China. Thus

[8] Confidential interviews with representatives of Chinese NGOs D and F, 16 and 21 January 2008, Beijing.

[9] Confidential interviews with representatives of Chinese NGOs D and F, 16 and 21 January 2008, Beijing.

[10] Confidential interview with representative of Chinese NGO K, 1 February 2008, Beijing.

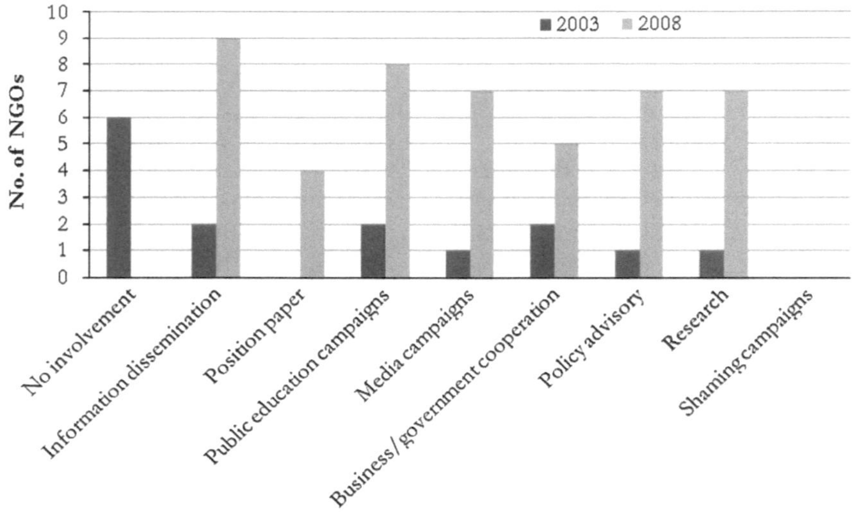

Figure 3. Climate-related campaigns
Source: Interviews with 8 Chinese NGOs and 4 INGOs with China offices, of which 6 were already operating in 2003. Beijing/Yunnan/Gansu: April 2007 and December 2007–February 2008

funding is quite a substantial form of transnational NGO cooperation, since only limited funding is available from Chinese individuals, for example FON stated that 58 per cent of their funding came from foreign sources in 2006. Other forms of cooperation are joint campaigns, publications, statements and exchange visits.

One landmark so far in transnational NGO cooperation has been the first joint position paper 'A warming China: thoughts and actions for the Chinese civil society' (FON et al 2007) at the Bali COP/MOP 13 in December 2007. Four Chinese and four international NGOs organized this survey based on responses by over 40 Chinese NGOs. The joint position taken, which was introduced at the Oxfam International side event at the COP/MOP 13, is conservative and in accordance with the Chinese government's position. Chinese NGOs regard this position paper as a first step, although some consider it 'boring', 'unsubstantial' and remark that 'no Western NGO would have published such a position. There is nothing to advocate about, because it is realized already'.[11] In contrast to the assumptions of the spiral model, one interviewee said that it was actually INGOs that advised them to take a moderate stand instead of raising points that go beyond the government's position. Another issue raised in interviews is the present lack of knowledge on climate change among most Chinese NGOs, which are not taking up energy- or climate-related campaigns themselves. This existing lack of knowledge and capacity is especially significant in relation to the Clean Development Mechanism (CDM) of the Kyoto Protocol. Despite the booming Chinese CDM market, only four (three INGOs and one Chinese NGO) out of 12 interviewed NGOs have CDM-related activities. These NGOs are active in CDM project development, provide CDM capacity development, and the WWF has launched a 'Go for Gold' campaign in which it asks athletes and spectators of the

[11] Confidential interviews with representatives of Chinese NGOs D and K and international NGO E, 16 January 2008 and 1 February 2008, Beijing.

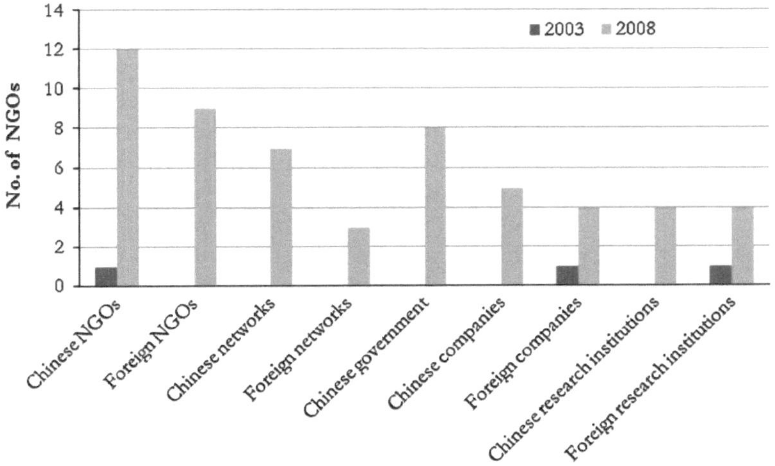

Figure 4. Cooperating partners for climate-related campaigns in 2003 and 2008
Source: Interviews with 8 Chinese NGOs and 4 INGOs with China offices, of which 6 were already operating in 2003. Beijing/Yunnan/Gansu: April 2007 and December 2007–February 2008

Beijing Olympics 2008 to offset their flights by Gold Standard CDM projects. When asked about their preferred partners for climate-related campaigns, NGOs in China mainly look to local and international NGOs, and to some extent also for governmental counterparts (see Figure 4).

Cooperation with international climate networks is in an initial stage. The Bali COP/MOP 13 was the first time Chinese NGOs participated in an international climate conference. Besides providing a great learning opportunity, Chinese NGO representatives could also get in touch with climate networks—such as the Climate Action Network (CAN) or Youth-based climate networks—and exchange ideas with the official Chinese delegation. Seven out of 12 interviewed organizations stated that participation in international conferences is a good method of getting international exposure, which in turn might increase their national reputation and status *vis-à-vis* the government. For example, one interviewee stated,

> Although we do not have much influence within the country, if we go to international meetings, we represent China, so it can influence their [the government's] position, what we do think. So the government is interested in what we think.[12]

Joining international networks on climate change is also seen as an opportunity to tap their vast knowledge on the issue.

> If NGOs are a network, then they will be more influential, then you can put all your intelligence together and come up with something. For an individual NGO it is difficult to catch up, they do not have enough capacity. Within the CAN network, you can even borrow international intelligence, ask your international partners, it is a bridge.[13]

---

[12] Confidential interview with representative of international NGO E, 16 January 2008, Beijing.

[13] Confidential interview with representative of Chinese NGO J, 24 January 2008, Beijing.

Grasping this opportunity to gain voice, two networks are currently forming in China: one will be CAN China, a network of Chinese NGOs active in climate-related campaigns; the other is China Youth CAN (CYCAN), a joint initiative of seven green student organizations. Although these two networks also have connections to international networks, they see their main focus of work at the national and even local level, CYCAN, for example, wants to initiate carbon inventories in 24 Chinese universities.

Asked about advantages and disadvantages of transnational NGO cooperation, most interviewees stressed funding, information exchange, mutual learning and easy access to expertise on climate change via their international partners as advantages, while three interviewees experienced 'cultural differences' and 'lack of understanding of the Chinese situation' as disadvantages. 'Mutual learning' has been named as one great expectation for transnational cooperation as Chinese NGOs begin to tap into the expertise of INGOs, while they also hope to convey information about the environmental situation in China—which is of better quality than information provided by the media—to INGOs and the western public. Eventually the exchange of knowledge and capacity appears to only be successful when happening at the right pace and on an equal footing: Chinese NGOs have stated to have felt pushed by INGOs towards ideas they did not fully support (for example how to set up joint networks) and also to have been hampered by INGOs, which displayed far more conservative points of view than the Chinese NGOs themselves (for example for joint declarations). Another issue for debate seems to be the focus of campaigns: INGOs are said to be targeting national and broad policy changes right away, whereas Chinese NGOs give priority to local projects for piloting policy changes.[14]

When asked to evaluate their campaign strategies for effectiveness on policy change, NGOs in China rated media and information sharing campaigns as the best way to make their positions heard by policy-makers (compare Figure 5). In most cases, this included organizing conferences or workshops for the dissemination of research results to which government officials were invited (see also Wexler et al 2006, 26). Depending on the sensitive nature of the research, these were either held internally, or with the participation of researchers, other NGOs, journalists and/or the interested public. Although their communication channels with decision-makers were not officially recognized, many NGOs had informal contacts (for example they have mobile numbers of their governmental counterparts working on climate change) and exchanged views via emails. All NGOs agreed that western NGO-style shaming strategies were not an option for them. Instead of critical campaigns, they preferred to give out positive incentives for behavioural change, for example assigning awards to city or local level governments which have made exceptional progress in energy saving measures.[15]

Overall, the Chinese NGO community (to the extent that it is represented in the interviews) is modest but optimistic about its impact on climate-related policies. Most interviewees observe an increasingly positive acknowledgement of their work by the Chinese government, and they also state that the major influence on

[14] Confidential interview with representative of Chinese NGO G, 21 January 2008, Beijing.

[15] Confidential interviews with representatives of Chinese NGOs F and K, 21 January 2008 and 1 February 2008, Beijing.

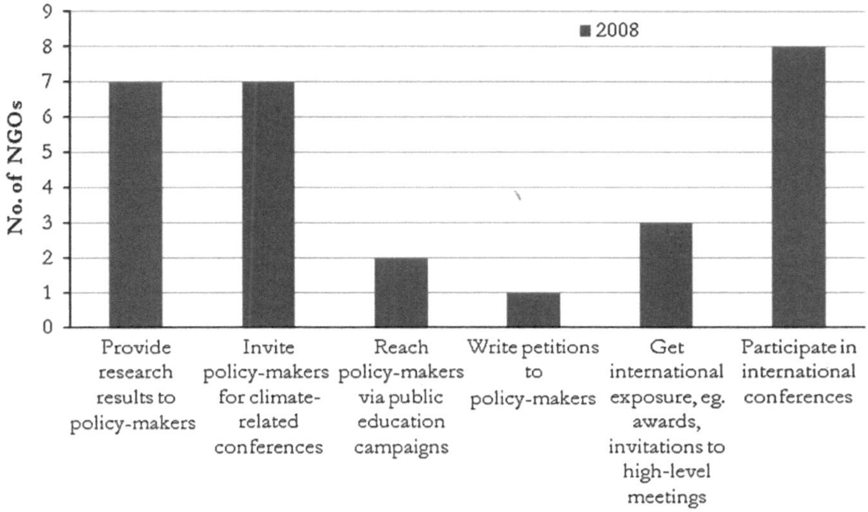

Figure 5. Strategies to influence climate-related policies
Source: Interviews with 8 Chinese NGOs and 4 INGOs with China offices, of which 6 were already operating in 2003. Beijing/Yunnan/Gansu: April 2007 and December 2007–February 2008

climate policies comes from Chinese research institutions. Major players among the Chinese research institutions that specialize on climate-related policy research, like the Chinese Academy of Social Science and the Energy Research Institute, can be considered as governmental think tanks with direct and official communication channels to political decision-makers (see also Heggelund 2007, 189).[16]

Discussion of results

The process analysis of Chinese climate politics tested the assumptions of the spiral model of norm internalization for the new issue area of climate politics. Using the case study of China, several assumptions could be verified: indeed, the change of Chinese climate politics happened gradually and can be conceptualized in different phases. The hypotheses of the spiral model on factors that trigger a transition from one phase to another could only be partly verified.

The influence of the transnational advocacy networks, which has the dominant role in pressuring for change in the spiral model, seemed to have less influence in the case study on climate change. Although some vulnerability of the target state could be identified for the 1990s, when China suffered from the loss of international reputation after the Tiananmen incident, pressuring occurred not until the mid-2000s, when representatives of OECD-governments started to address the issue of climate change regularly during state visits. Instead of pressuring tactics by a transnational advocacy network, the framing of the issue seemed to be important: climate change was denied as long as it was seen as an attribute of industrialized countries, but turned into a relevant topic once its impacts to China and its linkage

[16]Confidential interviews with representatives of international NGO E and Chinese NGOs F, G and J, 16 January 2008, 21 February 2008 and 24 February 2008, Beijing.

to alarming domestic environmental degradation was realized. However, the conveyers of the message, the 'norm entrepreneurs' in constructivist terms, do not seem to have been the transnational advocacy network of like-minded NGOs, but rather scientists that shared information with policy-makers. An explanation for this difference in advocacy strategies of human rights and climate advocates appears to be the role of science. In the field of human rights, appropriateness of behaviour, but hardly knowledge, shape a norm. In the field of climate politics, climate change has become a social and political problem due to the cause–effect relationships detected by natural science. Thus one can expect epistemic communities such as the IPCC to play a larger role in climate politics than in the human rights field.

The norm internalization process itself has relied more on the diffusion of knowledge and arguments than on shaming strategies (similar observations on strategies used by Chinese NGOs were made by Wexler et al 2006). Interviews with Chinese NGOs revealed that shaming strategies are not regarded as an option for them, because they prefer to take a 'constructive role' in which they only voice criticism if they can also provide solutions and instead tend to give positive incentives for behavioural change. Economic and social vulnerability and the urge for improving one's reputation could be detected as factors motivating the Chinese government to switch to a position closer to the international dominant norm. The notion of a learning process might therefore be a more appropriate description of a situation in which the target state was, at least partly, voluntarily getting involved in dialogue and action about climate protection, especially after it learned that this would be in its national interest. 'Complex learning' would be a fitting concept to describe a learning process which took part in a situation of uncertainty in which the actor is neither sure about her own identity, nor values guiding her behaviour (further discussed in Finnemore and Sikkink 1998, 908).

Conclusion

This article employed a constructivist approach, using the spiral model to track steps of norm internalization. Although we were able to trace changes in how the norm of climate protection was first denied, then debated and ultimately incorporated into national policies, the independent variables—meaning the factors and actors catalysing this change—were different from those of the original case. Two of the fundamental assumptions of the spiral model could not be confirmed: the transnational advocacy network did not use pressuring tactics, nor have INGOs been the main actors leading to a behavioural change of the Chinese government. Instead of a transnational advocacy network using shaming and pressuring strategies, we detected information sharing as the main trigger for a shift in climate politics. In contrast to the assumptions of the spiral model it had been epistemic communities, not NGOs, who had been the main conveyers of information. The message itself—high costs due to the impact of climate change and a growing challenge to achieving energy security—might have been the ultimate trigger for a change in Chinese climate politics. Norm internalization seemed to have grown with the realization of the costs of climate change, so that neither a constructivist nor a rationalist explanatory approach alone, but only one that intertwines both is able to account for a shift in Chinese climate politics.

For the international community advocating a certain norm, the experiences made in Chinese climate politics seem to recommend that a change in behaviour guided by norms can be best attained if the target state is convinced and not forced into the internalization of the norm. But the power of persuasion has its limits. In the case of climate protection, the international as well as the Chinese cascade of the norm of climate protection has so far found its limits in the debate about the appropriate level of economic development. As long as measures for climate protection have positive effects on economic development, or at least pose no regret options, governments are keen to position themselves as climate protectors. Yet as soon as norms limit actions and have negative effects on the economy, enthusiasm for climate protection is severely restricted.

References

ACEF [All China Environment Federation] (2006) *Zhongguo Huanbao Minjian Zuzhi Xinxi Ku* [*Information list of China's environmental NGOs*] (Beijing: All China Environment Federation)

Beijing Review (1991) 'Beijing Declaration on Environment and Development', 34:27, 9–13

Brie, Maria and Hagen Pietzcker (2004) 'NGOs in China: die Entwicklung des dritten Sektors [NGOs in China: the development of the third sector]', Discussion Paper, Wissenschaftszentrum Berlin für Sozialforschung [Social Science Research Center Berlin], Berlin

Cai, Shouqiu (1997a) 'Lun Ke Chixu Fazhan Dui Woguo Fazhi Jianshe De Yingxiang—shang [Discussion of the influence of sustainable development on the establishment of China's legal system Part 1]', *Faxue Pinglun* [*Law Discussion*], 1, 15–21

Cai, Shouqiu (1997b) 'Lun Ke Chixu Fazhan Dui Woguo Fazhi Jianshe De Yingxiang—xia [Discussion of the influence of sustainable development on the establishment of China's legal system Part 2]', *Faxue Pinglun* [*Law Discussion*], 2, 8–13

CSCPRC [Committee on Scholarly Communication with the People's Republic of China (CSCPRC)] et al (1992) *China and global change: opportunities for collaboration* (Washington, DC: National Academy Press)

Duan, Changqun, Xueqing Yang and Fayuan Gao (2000) 'Fifty years of political ecology in mainland China: evolving responses of political affairs to eco-environmental issues', *Sinosphere*, 3:1, 6–13, <http://www.chinaenvironment.net/sino/sino7.pdf>, accessed 19 May 2008

Economy, Elizabeth (1997) 'Chinese policy-making and global climate change: two-front diplomacy and the international community' in Elizabeth Economy and Miranda Schreurs (eds) *The internalization of environmental protection* (Cambridge: Cambridge University Press), 19–41

Economy, Elizabeth (2006) 'Environmental governance: the emerging economic dimension', *Environmental Politics*, 15:2, 171–189

Economy, Elisabeth (2007) 'The great leap backward?', *Foreign Affairs*, 86:5, 38–59

Finnemore, Martha and Kathryn Sikkink (1998) 'International norm dynamics and political change', *International Organization*, 54:4, 889–917

FON et al (2007) *A warming China: thoughts and actions for the Chinese civil society* (Beijing: Friends of Nature)

Foreign Ministry of the PRC (2007) 'Press briefing on Dec. 18', <http://english.gov.cn/2007-12/21/content_840424.htm>, accessed 13 June 2008

Haas, Peter M (1992) 'Knowledge, power, and international policy coordination', *International Organization*, 46:1, 1–35

Hardin, Garrett (1968) 'The tragedy of the commons', *Science*, 162:3859, 1243–1248

Hatch, Michael T (2001) 'Domestic politics and international negotiations: the politics of climate change in China', Paper prepared for the 2001 Berlin Conference on the Human Dimensions of Global Environmental Change, 7–8 December
Heggelund, Gorild (2007) 'China's climate change policy: domestic and international developments', *Asian Perspective*, 31:2, 155–191
Ho, Peter (2007) 'Embedded activism and political change in a semiauthoritarian context', *China Information*, 21:2, 187–209
Ho, Peter and Eduard B Vermeer (eds) (2006) *China's limits to growth: prospects for greening state and society* (Oxford: Blackwell Publishers)
IEA (2007) *World energy outlook* (Paris: International Energy Agency)
Johnston, Alastair I (1998) 'China and international environmental institutions: a decision rule analysis' in Michael McElroy, Chris Nielsen and Peter Lydon (eds) *Energizing China: reconciling environmental protection and economic growth* (Cambridge, Massachusetts: Harvard University Committee on Environment), 555–599
Keck, Margaret and Kathryn Sikkink (1998) *Activists beyond borders: advocacy networks in international politics* (Ithaca, New York: Cornell University Press)
Kuhn, Berthold (2006) 'In search of acceptable cooperation', *Development and Cooperation*, 33:7, 296–297
Levine, Mark D (1999) 'Opportunities for partnerships between the United States and China in energy efficiency research, development, demonstration and deployment', *Sinosphere*, 2:2, 4–11
Liu, Huaxin (1995) 'Quanqiu Hezuo Baohu Qihou [Global cooperation for climate protection]', *Renmin Ribao [People's Daily]*, 10 April, 7
March, James G and Johan P Olsen (1989) *Rediscovering institutions: the organizational basis of politics* (New York: Free Press)
Morton, Katherine (2007) 'Transnational advocacy at the grassroots in China: potential benefits and risks' in Peter Ho and Richards L Edmonds (eds) *China's embedded activism* (London: Routledge), 195–215
NBS [National Bureau of Statistics] (2006) *Green national accounting study report 2004* (Beijing: National Bureau of Statistics and State Environmental Protection Administration of China)
Netherlands Environmental Assessment Agency (2008) 'China contributing two thirds to increase in CO_2 emissions', Press release 13 June 2008, Den Hague, The Netherlands, <http://www.mnp.nl/en/service/pressreleases/2008/20080613Chinacontributingtwothirdstoincreasein CO2emissions.html>, accessed 25 August 2008
People's Daily Online (2006) 'Renmin zhishu shouci jieshe huanbao minyi jichu [The People's Index display the public opinion on the environment for the first time]', 23 January, <http://env.people.com.cn/GB/42691/4053264.html>, accessed 14 January 2008
Risse, Thomas, Stephen C Ropp and Kathryn Sikkink (eds) (1999) *The power of human rights: international norms and domestic change* (Cambridge: Cambridge University Press)
Risse-Kappen, Thomas (ed) (1995) *Bringing transnational relations back in: non-state actors, domestic structures and international institutions* (Cambridge: Cambridge University Press)
Ross, Lester (1988) *Environmental policy in China* (Bloomington: Indiana University Press)
Stokke, Olav S (1998) 'Understanding the formation of international environmental regimes: the discursive challenge' in Arild Underdal (ed) *The politics of environmental management* (The Netherlands: Kluwer Academic Publishers), 129–148
Turner, Jennifer L and Ellis Linden (2007) 'China's growing ecological footprint', *The China Monitor*, 16, 7–11
Wen, Bo (1998) 'Greening the Chinese media', in *China Environment Series*, 2 (Washington, DC: Woodrow Wilson International Center), 39–45
Wexler, Robyn, Ying Xin and Nick Young (2006) *NGO advocacy in China* (Beijing: China Development Brief)
Xiao, Zijian (1995) 'Kunjing he Jueze—Guanyu Renlei Xia Yi Ge Wenming He Fazhan Zhanlüe de Sikao [Difficult choice—thoughts on the next generation and the development strategy of human kind]', *Ziran Bianzhengfa Yanjiu [Studies in Dialectics of Nature]*, 2, 43–46

Yang, Guobin (2005) 'Environmental NGOs and institutional dynamics in China', *The China Quarterly*, 181:3, 46–66

Zhang, Guobao (2005) 'Speech at Minister forum of Beijing international renewable energy conference', 7 November, Beijing

Zhao, Liqing (1998) 'Dui Lianheguo Ke Chixu Fazhan Sixiang De Lijie [Understanding the UN principles on sustainable development]', *Shijie Jingji Yu Zhengzhi* [*World Economy and Politics*], 1, 36–38

Is climate change changing the EU? The second image reversed in climate politics

Oriol Costa[1]

Introduction

The participation of the European Union (EU) in international climate negotiations has attracted a good deal of scholarly attention. Climate change has become an important expression of the EU's 'actorness' in world politics (Bretherton and Vogler 2006); accordingly European foreign climate policy has been thoroughly examined (Harris 2007; Andresen and Agrawala 2002; Jaeger et al 1997; Schreurs 2004). However, the relationship between the EU and the international climate regime[2] (ICR) is not a one-way street—while the Union has been actively shaping the ICR, the latter has also had an effect on the EU. As has been elucidated by institutionalists studying 'second image reversed' processes

[1] This paper is part of a two-year study financed by a postdoctoral grant (2006 BP-A 10062), awarded by the research agency of the Catalan Government (AGAUR, Generalitat de Catalunya). I am grateful to Aron Buzógany, Vera van Hüllen, Erik Meier and four anonymous referees for insightful and helpful comments on earlier versions of this article.

[2] The most common definition of international regimes claims that they are 'principles, norms, rules, and decision making procedures around which actor expectations converge in a given issue-area' (Krasner 1983, 2). Accordingly, the international climate regime comprises of the UNFCCC, the Kyoto Protocol and the agreements reached by parties at the annual Conferences of Parties (COPs). The COPs themselves (and the rest of specialized bodies) are part of the regime, as far as they embody decision-making procedures.

(Gourevitch 1978), international negotiations and agreements can have a domestic influence (Kelley 2004).

Both bottom-up and top-down processes are present. For instance, the EU's stance in international negotiations has been shaped by member states and EU institutions. In turn, the EU has played a major role in building the ICR. When looking at the whole picture of the interaction, this uploading dynamics model bears a good deal of explanatory capacity. Thus it is only natural that several analytical approaches have been discussed in this regard, including regulatory competition (Héritier et al 1996; Weale 1996), the analysis of leadership (Gupta and Grubb 2000) and normative power (Lightfoot and Burchell 2004; Scheipers and Sicurelli 2007). However, this article argues that one needs to pay attention to how and when the ICR, as well as the negotiations and norms linked to it, have affected the EU's internal negotiations and decision-making processes. More specifically, accounting for downloading processes offers a way out of a central puzzle of EU climate policy: how has the EU—contrary to predictions from other scholars (such as Oberthür 1999, 655) and the logic of the 1971 European Road Transport Agreement (ERTA) ruling—managed to conduct an effective and ambitious foreign climate policy before it was able to harmonize the interests of states and adopt domestic policies?

The article proceeds as follows. The following section proposes a framework to analyse how and under what conditions international regimes have an impact on the EU. This framework is then applied to the EU's process of incorporating ICR norms and policies over the past two decades. The article relies on the existing literature as well as primary sources to trace processes and identify the international influence on key EU decisions regarding climate policies. Three phases are covered. The first phase (1985–1995) deals with the negotiations that took place before the adoption and entry into force of the United Nations Framework Convention on Climate Change (UNFCCC). During these years, the first international norm regarding climate change emerged, namely that developed countries must reduce their greenhouse gas (GHG) emissions according to multilateral agreements. The second phase (1995–2001) spans from the beginnings of the talks about the Kyoto Protocol (1997) to the adoption of the Bonn/Marrakech agreements (2001). The agreement on a second ICR norm—that market mechanisms have a key role in reducing emissions—takes place in this period. Finally, the third phase (2001–2008) begins with the announcement by George W Bush that the United States (US) did not intend to ratify the Kyoto Protocol and leads up to the present. The last section takes stock of the discussion presented and draws some conclusions.

When and how international regimes influence the EU

The claim that international regimes can have an impact on the EU should be unproblematic. After all, some International Relations (IR) scholars study how and under what conditions international institutions influence domestic politics (Checkel 1997; Cortell and Davis 1996; Dai 2007; Risse-Kappen 1995) and other academics in the field of European integration are also now addressing how the EU transforms its member states (Börzel 2005; Sedelmeier 2006; Radaelli 2003). Thus, given that the EU participates in a number of international negotiations and regimes, why should it not be conditioned by the policies and norms derived from the latter? Both sets of literatures—from IR and the field of European integration—provide a significant conceptual toolkit for the purposes of this article. However, some of their standard propositions need to be adjusted when dealing with the impact of international regimes on the EU. For example, these scholars usually grant a major role to the domestic conditions of the 'target' state. Accordingly, one expects that the specificity of the EU's governance structure (Héritier 1999) has some effect on the forms and scope of the influence exerted by regimes.

The first-cut arguments (and their limits)

International regimes influence internal political processes 'by way of the action of domestic political actors' (Cortell and Davis 1996, 451). Thus the first step to influencing the EU depends crucially on gaining the support of an agent (Dai 2007, 138). In the EU there are plenty of opportunities, as it 'provides an unusual abundance of access points to the policy-making process for interested actors' (Jönsson et al 1998, 328). The second step is more decisive. The domestic policy entrepreneur endorsing the regime must be successful in promoting the policies and norms of the regime within the Union.

These entrepreneurs can be regarded as 'change agents' (Börzel and Risse 2000, 9) that push for the incorporation of norms and policies associated with the regime. The question is whether or not the proposed change will garner the support of the actors that have veto power. The bibliography on veto players (Tsebelis 2002) provides some initial suggestions: the higher the number of veto players, and the bigger their differences about the proposed changes, the lower the probability of an entrepreneur being successful. Veto players also need to see that the proposed norms and policies are better than the status quo.

In the EU, given the diversity of decision-making procedures, the list of veto players varies. Thus, the Council of the European Union might stand as a single collective veto player or a sum of individual ones depending on whether it decides unanimously or by qualified majority. This, in turn, affects the clout of the European Commission, itself a veto player when it comes to Community matters. Likewise, the role of the European Parliament (EP) depends on whether the matter is under the co-decision or consultation mechanisms. There are also actors that have a *de facto*, instead of a *de jure*, capacity to veto (nonconstitutional veto players). These actors include certain Directorates-Generals (DG) of the Commission and the working parties of the Council. In sum, the policy entrepreneurs bolstering the regime need to operate within a system without too many veto players whose positions are not too different, and their proposal must be regarded as better than the status quo.

This argument is, nevertheless, an oversimplification. The indicated parameters depend on factors that might be considered as fixed, typically by constitutional dispositions (Tsebelis 2002, 2), but the key variables are relative in nature. For instance, how many veto players is too many? At what point do their differences become too wide? How are the policies and norms linked to a particular regime assessed? Veto players do not operate in a vacuum, but within a setting occupied by other actors that promote certain interests and approaches, construct coalitions and influence 'the domestic salience or legitimacy of [norms]' (Cortell and Davis 2000, 66).

In the EU, these factors have specific features. Two arguments regarding these features are advanced. The first pertains to the consequences that the fragmentation and lack of hierarchy of the EU policy-making system have on the influence of international regimes on the EU. The second deals with the fact that the scope of action of the EU is neither consolidated nor closed.

The fragmented-system argument

The EU offers numerous access points to decision-making, thus making it easier for international regimes to influence the agenda. At least two factors account for this. First, there is the relative lack of 'formal institutions which can adequately manage the policy dialogue' and facilitate the bargaining between actors (Wallace 1997, 10). Most notable among these absent 'off-the-shelf' institutions are political parties and party government (Peterson 1995, 69; Peters 1994, 15). Moreover, the presence of dense and diverse networks is related to 'the specialization of the EU as a political system producing regulatory policy' (Radaelli 1999, 759). This specialization favours the participation of actors that can offer technocratic or scientific expertise such as advisory boards and epistemic communities (Peterson and Bomberg 1999, 23). However, this is only one side of the coin. As Anthony R Zito argues, 'the EU system is a chain of institutions' that acts not only as *access* points, but also as *veto* points (2001, 586). Therefore, the construction of coalitions to modify the status quo complicates decision-making.

In the EU some conditions can facilitate the assembly of coalitions. These coalitions are frequently forged at the 'meso-level' of the EU (Peterson 1995), among and within the myriad of Council working groups, committees, Directorates of the Commission and so forth—which are generally run by specialized, middle-ranking policy-makers and bureaucrats. At this level, technocratic rationality tends to dominate; these actors approach decisions technically even when they are of a political nature. This approach frames the assessment of proposals, triggers self-restraint and shapes debates in terms of joint problem-solving (Lewis 2005, 943). Furthermore, despite belonging to different member states or institutions, similar functional agencies might still share the desire to push comparable policies forward (Peterson 1995, 78). The stability of the composition of these bodies, as well as the shadow of the future, facilitates processes of socialization and learning among its members (Héritier 1996, 156).

Naturally, not all issues are equally suitable for 'cooperative transgovernmental behaviour' (Peterson 1995, 78). In the words of John Peterson, history-making decisions[3] are shaped by very different actors, arenas and logics. These 'choices preoccupy the highest political levels in Europe' (Peterson 2001, 294) such as national cabinets and prime ministers meeting in Intergovernmental Conferences and are driven by political and legalistic rationalities. In reaching these decisions, the states are less exposed to potential transformative processes. Thus, one should expect less influence of regimes in these issues.

The European Commission plays a central role in the construction of coalitions. Networks 'normally coalesce around the Commission' and the

[3] Following Peterson, we use the expression 'history-making decisions' as a 'shorthand for the grand bargains that determine how and how much the EU changes' (2001, 294).

Directorates-General tends to 'build up specialized networks' around them (Jönsson et al 1998, 328), in the quest for resources that might enable them to promote their own agendas (Héritier 1999, 23). Beate Kohler-Koch points out that 'by organizing institutional platforms for exchange, selecting and bringing together experts', the Commission stimulates debates and gives them 'direction' in its own favour (2002, 4). It is unsurprising that the Commission exerts such influence. For instance, throughout the entire decision-making process, policy entrepreneurs have to vie with 'rival social networks'. As such, the Commissioner in charge of the dossier is 'the most obvious candidate' for such a job (Peterson and Bomberg 1999, 24). In short, the presence of the Commission in a coalition of policy entrepreneurs is vital if it is to be successful (Héritier 1996, 152), even beyond the observation that the Commission has the monopoly of legislative initiative.

The system-in-the-making argument

The forging of a coalition that brings together all the veto players depends on other variables. Entrepreneurs need to persuade other actors to back the norms and policies related to the regime, and 'build a collective good process that attracts supporters ... foster legitimacy to the case which binds the coalition and deflect external opposition' (Zito 2001, 586–587). All this depends on the coincidence of interests, approaches and the assessment that veto players make of the regime. The following paragraphs explore two implications of the fact that the EU is a political system *in the making*.

The first implication is related to the organizational consequences of policies: entrepreneurs must 'convince decision makers that a problem and a ready solution exist that fit the policy-makers' interest' (Zito 2001, 587; Peters 1994, 10). In the EU this has a specific implication. Given that 'the range of policies within the proper purview of the [EU] is not clearly defined', an entrepreneur 'may be able to expand the range of issues under consideration and with it expand the scope of Community action' (Peters 1994, 20; Héritier 1999, 8). In other words, the regime-related norms and policies that promise an increase in policy powers for the EU tend to find supporters more easily among those who favour a greater degree of European integration. The same argument applies to policies that help build the international actorness of the EU, or to policies that provide opportunities to empower specific EU institutions (Peterson 2001, 303).

The second implication of the EU being an open-ended, unfinished polity (Héritier 1999, 7) is that the adoption of regime-derived policies might require the creation of a new policy realm, rather than the mere modification of a pre-existing policy. As stated earlier, a number of factors influence actors' assessments of whether to change the status quo. Even if the decision-making mechanism is the same in a number of situations, a regime does not have, *ceteris paribus*, the same degree of success in influencing norms and policies. For instance, when policy entrepreneurs have to replace an adopted policy, the 'sunk costs' lower their chances of success (Peterson and Bomberg 1999, 19–20). In contrast, when the regime touches an unprecedented issue area, the absence of well-articulated interests and of 'strong preconceived views and beliefs' enhances the probability that a given actor would redefine its position and adopt an external rule (Bukovansky 1997, 232). Thus, the probability that a regime will influence the EU increases if the former precedes the existence of EU policies on the same issue.

Altering the 'domestic balance'

The influence that international regimes have on the EU depends largely on conditions related to its governance structure. However, these conditions may be influenced by the regime itself. By modifying the interests/preferences or the influence of some domestic actors (including veto players), regimes can alter 'the domestic balance' (Dai 2005, 388) and thus affect the chances of success for the policy entrepreneurs endorsing them.

International regimes can influence the actors' objectives either by changing their cost-benefit calculations or by modifying their underlying preferences.[4] Conditional membership is one of the clearest examples of the first possibility (Kelley 2004, 428), although sometimes membership is not made conditional but the establishment of privileged relations (Sedelmeier 2006). The shadow of the future and reciprocity may also alter the choices of some actors. To be sure, the rewards and punishments might be more ideational than material and might be linked to reputation, shaming and praising. Moreover, if one relaxes the rationality assumption, the regime-derived norms or policies can provide rough templates should actors prefer to adopt solely satisfactory but ready-made strategies (Cortell and Davis 1996, 453). Finally, sociological and constructivist approaches have maintained that regimes can not only modify actors' strategies, but also their underlying preferences, through 'processes such as influence or persuasion' (Kelley 2004, 428) and 'social learning' (Börzel and Risse 2000, 7).

In short, regimes may shape the goals and behaviour of actors, thus making it easier or more difficult for regime-oriented entrepreneurs to put their argument across. Moreover, regimes can also influence the distribution of power between domestic actors and therefore constrain or facilitate the expression of certain interests or ideas. On occasions, regimes enable some actors to access decision-making processes that would have been out of their reach otherwise (Abbott and Snidal 2000, 428). In addition, they can 'increase the political leverage and further improve the informational status of pro-compliance constituents' (Dai 2007, 8). Thus, Checkel argues that domestic actors use international norms 'to generate pressures … on state decision-makers' (2001, 557). Finally, regimes also help to legitimize certain proposals or policies, thereby shifting the burden of proof onto other actors.[5]

The capacity of international regimes to alter the domestic balance depends on their strength, understood in terms of both the stringency of their constraints as well as the robustness and endurance of their institutional setting. In other words, a regime has more domestic influence when it 'significantly constrains the range

---

[4] The question of how and why international regimes matter has been addressed comprehensively by Levy et al (1995). This article only addresses the influence of regimes on domestic policy-making processes.

[5] There are some synergies between the approach presented here and that of scholars working on institutional interaction. Oberthür and Gehring have identified four different causal mechanisms of institutional interaction: 'cognitive interaction', 'interaction through commitment', 'behavioural interaction' and 'impact-level interaction' (2006). The first two mechanisms resonate with the discussion in this article, as they deal with international institutions changing the objectives and opportunities of actors in other settings. Furthermore, the empirical phenomena they are interested in overlap with the case study of this article (Oberthür 2006). However, these scholars look at how institutional interaction affects the effectiveness of international institutions, whereas this article focuses on the conditions under which the interaction is more influential upon the EU.

of behaviour that qualifies as legal or appropriate' (Underdal 2004, 29) and encourages actors 'to adopt extended times horizons' (Underdal 2002, 26). However, it must be stressed that it is not the legal force *per se* of international norms that is most relevant, but their capacity to alter the domestic correlation of forces. Thus 'even weak institutions and soft laws can impact national policies' (Dai 2005, 388). This is also the reason why a regime might be influential before the EU (or its member states) has decided to ratify the relevant international agreements, or even after it has explicitly refused to do so. Furthermore, this process can be accumulative, in the sense that if relevant actors consistently appeal to international rules, they become taken for granted to a certain degree (Cortell and Davis 2000, 81).

The influence of the international climate regime on the EU

This section builds on the argument previously outlined. It identifies the conditions that have enabled the ICR, and international climate negotiations, to influence the EU. It shows that some of these conditions have been favoured by the existence of the regime itself. This influence has altered the EU's internal negotiations and decision-making process, as well as European integration itself. As argued above, this process has spanned two decades, during which the regime has been strengthened in terms of its stringency. The ICR has progressively incorporated broad principles (UNFCCC), targets and timetables (Kyoto) and fine-grained rules as to how to meet them (Bonn/Marrakech). However, the evolution of its institutional robustness has been more mixed. On one hand, the entry into force of the UNFCCC and of the Kyoto Protocol, as well as the consolidation of the institutional procedures, has set up a strongly institutionalized regime. On the other hand, the stance taken by the US seems to threaten the endurance of the regime on occasions.

1985–1995: going along with states and international negotiations

By the end of the 1980s, 'a norm requiring developed countries to accept a domestic CO_2 emission reduction commitment was emerging', to the point that 'by 1990 the United States was almost completely isolated' in its rejection (Cass 2005, 42). The increasingly successful ozone negotiations loomed large on states, international organizations and public attention (Oberthür 1999, 658). Initially, the European Community (EC) approached climate change in a reactive fashion. However, two mutually reinforcing dynamics changed the picture: first, the establishment of the Intergovernmental Panel on Climate Change (1988) and the holding of international conferences (Villach 1985 and 1987, Toronto 1988, The Hague 1989, Noordwijk 1989, Geneva 1990), and second the initiatives of particular member states (the first ICR-oriented policy entrepreneurs in the EC) (Grubb 1995, 43).

The Netherlands was especially proactive. Climate change had already appeared in its first 'Indicative Multi-year Programme for the Environment' (1985) and in 1989 The Netherlands hosted the Ministerial Conference on Atmospheric Pollution and Climatic Change (Noordwijk) (Kanie 2007, 92). In Germany, climate reached the agenda in the context of the post-Chernobyl debate on energy. In 1987 the highly influential Inquiry Commission on Precautionary Measures for the Protection of the Earth's Atmosphere was established in the

Bundestag and in June 1990 the government announced its commitment to reducing CO_2 emissions by 25 per cent by 2005 (Cavender and Jäger 1993, 9). The United Kingdom also began addressing the issue in the late 1980s and, from April 1989 onwards, actively promoting the negotiation of a climate convention (Cass 2006, 23). By autumn 1990 a fair number of EC member states had already established objectives or plans for the reduction of emissions.

The EC waited until October 1990 to decide on its position. To be sure, in 1986 the EP had passed a resolution on climate, and two communications by the Commission (1988 and 1989) revised some scientific aspects of the affair and mentioned possible policies. However, the Fourth Environmental Action Programme (proposed by the Commission and covering the years 1987 to 1992) did not even mention global warming. Similarly, the Programme of the Commission for 1989 included only the vaguest of commitments: 'the Commission will make proposals with a view to enabling the Community to play a constructive role on a common basis' (European Commission 1989, 38). It was not until 1990, and in preparation for international negotiations (Michaelowa 1998, 153), that 'the process picked up considerable momentum' (Wettestad 2000, 27). The Dublin European Council of June 1990 decided to establish targets to limit emissions, presenting the following argument:

> The Community and its Member States have a special responsibility to encourage and participate in international action to combat global environmental problems. Their capacity to *provide leadership* in this field is enormous. (European Council 1990, 18, emphasis added)

Accordingly, in October, a joint Council of Energy and Environment Ministers adopted the objective of stabilizing CO_2 emissions in 2000 at 1990 levels, which became the negotiating position of the EC/EU for some years. The objective was surprisingly ambitious considering that the unanimity rule (which applied to environmental affairs until the Maastricht Treaty) multiplied veto players. The objective immediately became a baseline for both international negotiations and negotiations within the EC. However, some authors have qualified the audacity of the target, arguing that the decision was 'a reflection that the targets already declared by member countries ... would almost suffice to achieve 1990–2000 stabilization across the EU' (Grubb 1995, 43). Probably the best indication of the primacy of national objectives is that the acceptance of a shared target did not lead to the emergence of a common EC position in the negotiations for the UNFCCC (Vogler 1999, 31).

Two additional aspects should be stressed. First, the agreement was made before the Second World Climate Conference (Geneva, November 1990), and it was shaped by the EC's explicit will to assume leadership expressed at the Dublin Council and shared by the Commission (this is, shared by all the veto players). The overarching aspiration, according to Brian Wynne, was for the EC to have a 'major presence at the Rio Earth Summit' (1993, 201). In other words, climate negotiations were perceived as way 'to stand forth as a strong and unified block on the world scene' (Andresen and Agrawala 2002, 45). This reframed the internal debate and made the EC more prone to adopt an ambitious stance. Second, the decision to lead enabled the Environment DG to make a simple argument: if the EC wanted to maintain its international leadership role during the negotiations of the UNFCCC, it needed to adopt policies to control its GHG emissions (Wettestad 2000, 28; European Commission 1991). Thus, since 1993 the Commission reported annually to the EP and Council on whether the Community was making progress to ensure the stabilization of emissions, invariably reaching the conclusion that further European-wide measures were needed (Skjærseth 1994, 25). The Commission therefore became the key entrepreneur for the incorporation of the ICR norms as internal policies.

However, the content of the Commission's proposal left veto players divided (in particular, states). The first package of measures was made up of (1) a carbon and energy tax, (2) measures on renewables and energy efficiency and (3) a monitoring mechanism. The carbon/energy tax was the 'cornerstone' of the proposal (Haigh 1996, 163) but its controversial nature almost derailed the entire package. The introduction of the tax would have two major institutional consequences: the creation of a fiscal policy and an energy policy for the EC. The Commission's proposal alienated both actors who were opposed to ecological taxes and those who were opposed to these taxes being fixed and collected by the EC.[6] The proposal was also controversial within the Commission, where the commissioner of the DG XXI (taxation) 'opposed the tax from the outset' (Skjærseth 1994, 28). Hence, the package of measures was only adopted in the form of a 'watered down version' in 1993 (Wettestad 2000, 28). In other words, the ICR made it to the agenda of the EU, but failed to significantly alter its internal policies.

Consequently, the EC attended the Rio Summit in the midst of a considerable internal controversy (Bretherton and Vogler 2006, 13). However, the active participation of some of its member states (in particular, Denmark, Germany and The Netherlands) and the reluctant attitude of the US enabled the Community to continue exercising its role as the *pusher* of climate negotiations. Moreover, the will to lead and the stabilization objective (both associated with international negotiations) became indisputable principles in EU debates (Grubb 1995, 43) and led to the emergence of a more solid position over the following years. Despite its own weakness, the negotiation of the ICR encouraged the building up of a common EU approach.

1995–2001: international ambition and domestic scantiness

After the UNFCCC (1994) came into force, international negotiations centred on the preparation of the Kyoto Protocol and (from 1998 onwards) on the rules for its implementation (Bonn/Marrakech agreements). During this period the regime became stronger, both institutionally and in terms of its stringency, locking in decisions and decision-making procedures that favoured an ambitious EU stance. In addition, some internal developments made the EU more amenable to the influence of the ICR. As a result, the EU boldly incorporated the main norm derived from the ICR—that developed states have to reduce their emissions and do so in a cooperative fashion. Nonetheless, internal policies continued to be essentially unchanged, although Kyoto helped trigger a change that would bear fruit in the following stage.

The EU failed to translate the objective of controlling GHG into domestic policies. Maastricht had introduced qualified majority voting for environmental

---

[6] The UK was particularly opposed to the EC having the power to fix and collect taxes.

affairs, reducing the number of veto players. However, the Commission continued to push for the tax proposal, which returned the decision to the rule of unanimity[7] and negatively conditioned the position of states regarding a common climate policy. European climate policy was 'in a delicate situation' (Jaeger et al 1997, 196). Actually, the most outstanding progress—the burden-sharing agreements of 1997 and 1998—was closely related to international negotiations. In 1991 the Commission had already floated the idea of sharing the common target, but the United Kingdom, France and Italy opposed it. But after the first Conference of Parties (COP1) (1995), 'with the credibility of the EU on the global stage at the forefront of their minds', the ministers reached an agreement on a target and on a way of sharing it (McCormick 2001, 285). Thus, international negotiations encouraged member states to reach an agreement (Skjærseth and Wettestad 2008, 5), which in turn brought the positions of veto players closer together, and facilitated later attempts to adopt domestic measures (Hovi et al 2003, 10; Oberthür 2006, 69).

On the contrary, the international dimension of climate policies was relatively uncontroversial, especially from Kyoto onwards. The EU achieved greater unity than in the previous period and defended the idea that the ICR should not be watered down to make compliance with the targets easier. Indeed, the differences between member states only became relevant towards the end of the period, when the US insisted on making the flexibility mechanisms more flexible. What is most interesting here is that the participation of the EU in the negotiation of the ICR brought about three new circumstances.

First, international negotiations empowered a new coalition of policy entrepreneurs, who have been in favour of the EU adopting ambitious international and domestic commitments. The Working Party on International Environmental Issues/Climate Change (WPIEI/CC), created after the COP1 (1995), is the main locus of this coalition and one of the expert groups assisting the Council. It is comprised of the heads of the climate units in environment ministries and its remit includes the formulation of the EU's stance before international negotiations, as well as the preparation of the global warming part of the Council conclusions (Costa forthcoming). Although the WPIEI/CC is only supposed to prepare the technical aspects of the ministries' decisions, it has in practice assumed a political role. That is, by representing the EU in international negotiations, the WPIEI/CC has acquired a say in the overall scope of the internal measures. Importantly, the WPIEI/CC is not simply an intergovernmental arena. As described for the COREPER and other working groups (Lewis 2005, 943), its members have a vested interest in promoting an ambitious approach to climate policy. An organizational culture seems to have emerged. The Environment DG is also part of this group of policy entrepreneurs and actually takes part in WPIEI/CC meetings. In fact, the relation between both bodies appears to be based more on collaboration than on institutional competition (Groenleer and van Schaik 2005, 13). Additionally, it must be stressed that the Environment Ministers usually give approval to the decisions taken by the WPIEI/CC, probably due to the coincidence of their interests and policy paradigms. In sum, the negotiation of the ICR empowered a cross-national

[7] Article 175.2 of the EC Treaty provides that environmental measures 'primarily of a fiscal nature' (among others) are to be adopted unanimously.

and cross-institutional alliance of middle-ranking entrepreneurs representing similar functional agencies (Peterson 1995, 78).

Secondly, Kyoto (and the reluctance of the US) 'provided the EU with a unique opportunity to capitalize on its economic and environmental presence and to assume a leadership role in the climate change regime' (Bretherton and Vogler 2006, 9). The explicit will to lead the negotiations legitimized policy entrepreneurs (Damro and Luaces Méndez 2003, 79) and attracted the endorsement of some actors who were concerned about the external relations of the EU. This is the case for the Presidency of the Commission, the DG of Exterior Relations and the European Parliament, which joined the WPIEI/CC and the Environment DG in the defence of an ambitious EU stance on climate change.

Finally, the inclusion of flexibility mechanisms in the Protocol meant the adoption of a second international climate norm: that market mechanisms have a central role to play in reducing GHG emissions. This induced a change within the EU. Months before the COP3, the EU had only accepted emissions trading as a concession to the US, in exchange for the latter's acceptance of binding targets. However, the EU attitude towards this compromise has changed from initially lukewarm to a more positive one. This change of attitude was a result of a number of factors including the inevitability of emissions trading, the will to play an important role in its operation, the pressing need to match the international ambition with internal policies, the slowness of the post-1997 negotiations and a learning process triggered by the Kyoto talks (Wettestad 2005; Skjærseth and Wettestad 2008).

2001–2008: crisis in the negotiations, domestic policies and unilateral commitments

In this final stage the EU notably reinforced the incorporation of norms derived from the ICR in both its international and domestic policies. This section highlights two significant processes which have strong links with the international climate negotiations and regime. The first is the reaction to the announcement that the US was not going to ratify Kyoto. This imperilled the viability of the ICR and thus could have led to a reduction of its influence on the EU. A regime that might not survive is certainly less able to encourage compliance. However, the EU has a vested interest in the survival and performance of the ICR and thus the weakness of the regime paradoxically increased its influence on the EU. The second process is the full assumption of emissions trading as one of the EU's internal policies.

In March 2001, George W Bush announced that the US had no intention of joining the Kyoto Protocol. Although the US' attitude had traditionally been one of reluctance, Bush's decision threatened the very existence of multilateral frameworks with reduction targets. The announcement had at least three paradoxical effects on the EU. First, it reinforced the coalition of policy entrepreneurs by increasing the political appeal of leading international climate negotiations, which had been languishing since 1997. Bush's announcement, and the position of Australia under John Howard, 'sparked off a storm of international criticism and united most of the world in opposition to the United States' attitude' (Vrolijk 2002, 3). Secondly, the US' withdrawal led to the EU reaching the conclusion that the ICR should remain even without US participation. This decision has been said to be 'historic', given that during the 1990s the EU had conditioned its commitments on the reciprocity of its main partners, meaning the US and Japan (Schreurs 2004, 209). Moreover, the EU has progressively been edging towards the adoption of unilateral binding targets, which indicates that the EU takes the norm for granted that developed states must reduce their GHG emissions.[8] Finally, the new status of the ICR modified the EU's priorities. Now it was a case of saving Kyoto rather than negotiating the rules for its implementation (an issue that had put European unity at risk in the COP6 in The Hague). So, if 'prior to the Bush announcement, cracks were evident' among veto players (particularly among states), the Bush decision 'so angered the Europeans that they were able to ... present a strong united block' (Schreurs 2004, 209). This new strategic environment in international climate negotiations brought together the positions of the key actors and accelerated the establishment of the EU Emission Trading Scheme (EU-ETS), overcoming the resistance of 'skeptical Environmental NGOs, Commission Directorates, Member States and Members of the European Parliament' (Wettestad 2005, 17).[9] In effect, during this period, the EU fully incorporated a norm derived from the ICR—that market mechanisms are an important instrument in the reduction of GHG—into its internal policies. After years of 'tax deadlocks and limited success' (Christiansen and Wettestad 2003, 6), emissions trading had become the 'jewel in the crown for EU climate policy' (Wettestad 2005, 17).

Two aspects of the EU-ETS seem particularly relevant here. First, it was 'Commission officials within the DG Environment' who promoted the trading option, from as early as 1998. One of the factors that accounted for this is that emissions trading was identified as an instrument that could 'exert control over the further EU climate policy process and pull the EU together' (Wettestad 2005, 12). Furthermore, some member states had started to consider establishing national trading systems, which was perceived as a threat to the internal market (Christiansen and Wettestad 2003, 7). Thus, the Commission altered 'its longstanding opposition to emissions trading' (Cass 2005, 48) and in 2003 the EU adopted the main directive for an emissions trading system with some 12,000 installations. Interestingly, throughout this process the Commission played an 'extraordinarily strong role', and its 2001 proposal was 'only marginally altered by the European Parliament and the Council of Ministers' (Wettestad 2005, 2). Second, the EU-ETS has eased the development of EU-domestic climate policies. Along with the EU's commitment to lead, the acceleration of the EU-internal policies has led to the adoption (or the proposal) of more comprehensive policies, such as the January 2008 proposal by the Commission. This 'climate and energy package' includes relevant energy measures, comprising binding targets of a 20 per cent share of renewables and 10 per cent for bio fuels, the harmonization of guarantees of origin and binding national targets in line with the overall EU ones (European Commission 2008). In short, the adoption of policies derived from the ICR is facilitating a certain expansion of the EU's scope of action, and this partially explains the behaviour of key institutional actors (Schreurs and Tiberghein 2007, 34).

---

[8] The Environment Council of February 2007 agreed to reduce the EU's emission by 30 per cent by 2020 if other developed countries adopted an equivalent commitment, or otherwise 20 per cent (Council of the European Union 2007a). In March, the European Council endorsed these objectives (Council of the European Union 2007b).

[9] Other factors played a role as well. See Skjaerseth and Wettestad (2008).

Conclusions

During the last two decades, the incorporation of ICR-related norms and policies into the EU has displayed two major features. On one hand, it has occurred gradually. Since 1990 there has been an increase in the degree of ambition and in the number of sectors in which the EU has adopted measures to reduce emissions. In this respect, the evolution has been rather remarkable from the Council agreement of October 1990 to the policies adopted or proposed over the last years. On the other hand, this development has occurred at two speeds and with two different intensities. In the international dimension, the EU has assumed a more ambitious attitude at an accelerated rate, whereas in the internal realm the adoption of ICR-related norms and policies has been slower. This development is counterintuitive. One would expect that to act as an international leader the EU would first need to acquire some internal competences and gather a basic consensus on specific domestic policies. This article claims that this development is due to the influence of the ICR on the EU. To a certain extent, the EU has constructed a climate policy on the basis of its participation in the ICR (Sjöstedt 1998, 231).

The incorporation of ICR-related norms and policies parallels the increasing strength of the regime, despite the fact that its institutional setting seemed rather fragile on occasions. Norms, targets and mechanisms have gradually been locked-in in the ICR, thus framing successive climate negotiations and in turn the EU stance. However, even when there were only climate *talks* and no commitment to control GHG emissions, international processes had a relevant influence on the EC/EU. Therefore the influence is better explained by also taking into account the conditions faced by policy entrepreneurs, which in turn have been shaped by the very existence of the international climate negotiations and regime.

Two arguments were advanced in this article. The first pointed out that the possibilities of a regime influencing the EU depend on the number of veto players, the distances between them and the assessment they make of the policies and norms derived from the regime; in the case of climate change these variables have improved over time. Maastricht established the qualified majority vote on environmental issues. The withdrawal of the tax proposal changed political venues and led to the disappearance of the veto on what could have become the centrepiece of domestic policies. Similarly, veto players have slowly formed around the policy entrepreneurs' proposals, first in the international dimension and later in the domestic one. The international negotiations and regime have played a role in this. They have legitimized the positions of policy entrepreneurs and made it more difficult for those who preferred to stick to the status quo. Certainly, 'European public opinion [has been] strongly behind Kyoto' (Schreurs and Tiberghien 2007, 30) and the Kyoto Protocol pretty much made the case for the EU-ETS. The ICR has also offered some political and institutional incentives, making an ambitious stance on climate change more appealing for several actors (see below).

The second argument proposed in the second section is related to the lack of hierarchy and the weak institutionalization of the EU. Entrepreneurs endorsing a regime have greater possibilities of success if the norms and policies they promote are suited to technocratic and bureaucratic rationality. More importantly, the entrepreneurs need the Commission's support. In this regard, climate policies have been placed in the 'meso-level'. In fact, only the tax proposal acquired the characteristics of a history-making decision. Therefore, alliances have emerged that have systematically favoured the incorporation of ICR-related norms and policies. These dynamics are best represented by the WPIEI/CC (and its relationship with the Environment DG), which has become an appropriate framework for cutting slack, promoting socialization and facilitating collaboration between similar functional agencies. The existence of the climate regime has empowered these entrepreneurs, if not encouraged its appearance, insofar as it has given them access to decision-making and legitimized their arguments. It is the participation of the EU in international climate negotiations that triggered the creation of the WPIEI/CC and has enabled it to play a central role. Here it must be recalled that the Commission is part of the core group of policy entrepreneurs. With the exception of the brief initial period in which the initiative pertained to some states, the Commission has sponsored all the initiatives to adopt ICR-related norms and policies. This is true both for the ICR norm advocating the adoption of binding commitments in multilateral settings and for that concerning the use of market instruments to reduce emissions. Although the Commission has experienced internal debates, the general trend has been clearly ICR-oriented. A good reason for this is that the rules derived from the ICR have enabled the Commission (and the EU) to expand its policymaking capacity.

This confirms the third argument presented in the second section, namely, that the nature of the EU as a political system *in the making* can facilitate the task of policy entrepreneurs. Climate change has been perceived as a way of both reinforcing the EU's powers and of developing policies in less propitious fields. As John McCormick argues, 'public opinion has been supportive of the role of the EU in environmental policy' and 'few doubts apparently remain in the minds of Europeans about the value of EU activity in the field of the environment' (2001, 61). Equally, it has also offered a way of promoting the EU as an international actor with the capacity for leadership. Environmental protection 'has become a major issue for EU foreign policy' (Schreurs 2004, 222), and the 'EU elites' have seen in this a way to 'increase public support for EU integration' (Schreurs and Tiberghien 2007, 26). In short, the institutional consequences of the European leadership of the ICR have contributed to the gathering of a winning coalition that goes well beyond environmental administration.

Therefore, the conditions have become more suitable for the climate regime to have an influence, partly because the ICR has been able to shape them. It has opened up opportunities that the Commission and environmental ministries have exploited to gain influence and autonomy. However, the study suggests yet another interaction. Although the ICR has triggered some degree of European integration, the EU has also increased the exposure of its member states to the regime. Due to its governance structure, the EU has acted as a resonator of international institutions. Miranda Schreurs and Yves Tiberghein have argued that EU climate policies are 'the result of a dynamic process of competitive multi-level reinforcement among the different EU political poles within a context of decentralized governance' (2007, 22). This article presented some specific examples of this kind of processes. The singularity of the EU as a polity has encouraged the succession of policy entrepreneurs, as well as the attempts by the Commission to regain leadership when it has been exercised by states, the use of climate change as an opportunity, the important role of the WPIEI/CC and its capacity to set the tone for domestic debate. The case of climate change seems to indicate that the EU has the unintended consequence of making its member states more open to the influence of regimes. Of course, it takes more than one case study to prove that this argument is valid generally. More empirical analysis is needed. However, further research does appear to offer promising avenues to explore the top-down relationships between international regimes and the EU.

References

Abbott, Kenneth and Duncan Snidal (2000) 'Hard and soft law in international governance', *International Organization*, 54:3, 421–456

Andresen, Steinar and Shardul Agrawala (2002) 'Leaders, pushers and laggards in the making of the climate regime', *Global Environmental Change*, 12:1, 41–51

Börzel, Tanja (2005) 'Europeanization: how the European Union interacts with its member states' in Simon Bulmer and Christian Lequesne (eds) *The member states of the European Union* (Oxford: Oxford University Press), 45–69

Börzel, Tanja and Thomas Risse (2000) 'When Europe hits home: Europeanization and domestic change', *European Integration Online Papers* (EIoP), 4:15, <http://eiop.or.at/eiop/texte/2000-015a.htm>, accessed 13 February 2008

Bretherton, Charlotte and John Vogler (2006) *The European Union as a global actor*, second edition (New York: Routledge)

Bukovansky, Mlada (1997) 'American identity and neutral rights from independence to the war of 1812', *International Organization*, 51:2, 209–243

Cass, Loren (2005) 'Norm entrapment and preference change: the evolution of the European Union position on international emissions trading', *Global Environmental Politics*, 5:2, 38–60

Cass, Loren (2006) *The failures of American and European climate policy: international norms, domestic politics and unachievable commitments* (New York: State University of New York Press)

Cavender, Jeannine and Jill Jäger (1993) 'The history of Germany's response to climate change', *International Environmental Affairs*, 5:1, 3–18

Checkel, Jeffrey (1997) 'International norms and domestic politics: bridging the rationalist-constructivist divide', *European Journal of International Relations*, 3:4, 473–495

Checkel, Jeffrey (2001) 'Why comply? Social learning and European identity change', *International Organization*, 55:3, 553–588

Christiansen, Atle C and Jørgen Wettestad (2003) 'The EU as a frontrunner on greenhouse gas emissions trading: how did it happen and will the EU succeed?', *Climate Policy*, 3:1, 3–18

Cortell, Andrew P and James W Davis, Jr (1996) 'How do international institutions matter? The domestic impact of international rules and norms', *International Studies Quarterly*, 40:4, 451–478

Cortell, Andrew P and James W Davis, Jr (2000) 'Understanding the domestic impact of international norms: a research agenda', *International Studies Review*, 2:1, 65–87

Costa, Oriol (forthcoming) 'Who decides EU foreign policy on climate change? Actors, alliances, and institutions' in Paul Harris (ed) *Climate change and foreign policy: case studies from East to West* (London: Routledge)

Council of the European Union (2007a) 'Press release: 2785th Council meeting. Environment', Brussels, 20 February, 6272/07 (Presse 25), <http://www.consilium.europa.eu/ueDocs/cms_Data/docs/pressData/en/envir/92864.pdf>, accessed 10 August 2008

Council of the European Union (2007b) 'Brussels European Council 8/9 March 2007: Presidency conclusions', Brussels, 2 May, 7224/1/07 REV 1, <http://register.consilium.europa.eu/pdf/en/07/st07/st07224-re01.en07.pdf>, accessed 10 August 2008

Dai, Xinyuan (2005) 'Why comply? The domestic constituency mechanism', *International Organization*, 59:2, 363–398

Dai, Xinyuan (2007) *International institutions and national policies* (Cambridge: Cambridge University Press)

Damro, Chad and Pilar Luaces Méndez (2003) 'Emissions trading at Kyoto: from EU resistance to Union innovation', *Environmental Politics*, 12:2, 71–94

European Commission (1989) 'Programme of the Commission for 1989. Address by Jacques Delors, President of the Commission, to the European Parliament and his reply to the debate', 16 February, <http://aei.pitt.edu/7631/>, accessed 15 March 2008

European Commission (1991) 'A Community strategy to limit carbon dioxide emissions and to improve energy efficiency', Brussels, SEC(91)1744 final, 14 October

European Commission (2008) '20 20 by 2020—Europe's climate change opportunity', Brussels, COM(2008)30 final, 23 January

European Council (1990) 'Dublin Summit', *Bulletin of the European Communities*, 6/1990, 25–26 June

Gourevitch, Peter (1978) 'The second image reversed: the international sources of domestic politics', *International Organization*, 32:4, 881–912

Groenleer, Martijn LP and Louise G van Schaik (2005) 'The EU as an "intergovernmental" actor in foreign politics: case studies of the International Criminal Court and the Kyoto Protocol', Working Document 228/August 2005, Centre for European Policy Studies, Brussels

Grubb, Michael (1995) 'European climate change policy in a global context' in Helge Ole Bergesen, Georg Parmann and Oystein B Thommessen (eds) *Green globe yearbook of international co-operation and development 1995* (Oxford: Oxford University Press), 41–50

Gupta, Joyeeta and Michael Grubb (2000) 'Climate change, leadership and the EU' in Joyeeta Gupta and Michael Grubb (eds) *Climate change and European leadership: a sustainable role for Europe?* (Dordrecht, The Netherlands: Kluwer Academic Publishers), 3–14

Haigh, Nigel (1996) 'Climate change policies and politics in the European Community' in Timothy O'Riordan and Jill Jäger (eds) *Politics of climate change: a European perspective* (New York: Routledge), 155–185

Harris, Paul (ed) (2007) *Europe and global climate change: politics, foreign policy and regional cooperation* (Cheltenham, United Kingdom: Edward Elgar)

Héritier, Adrienne (1996) 'The accommodation of diversity in European policymaking and its outcomes: regulatory policy as a patchwork', *Journal of European Public Policy*, 3:2, 149–167

Héritier, Adrienne (1999) *Policy-making and diversity in Europe: escape from deadlock* (Cambridge: Cambridge University Press)

Héritier, Adrienne, Christoph Knill and Susanne Mingers (1996) *Ringing the changes in Europe: regulatory competition and redefinition of the state. Britain, France, Germany* (Berlin: Walter de Gruyter)

Hovi, Jon, Tora Skodvin and Steinar Andresen (2003) 'The persistence of the Kyoto Protocol: why other Annex I countries move on without the United States', *Global Environmental Politics*, 3:4, 1–23

Jaeger, Carlo C, Terry Barker, Ottmar Edenhofer, Sylvie Faucheux, Jean-Charles Hourcade, Bernd Kasemir, Martin O'Connor, Martin Parry, Irene Peters, Jerry Ravetz and Jan Rotman (1997) 'Procedural leadership in climate policy: a European task', *Global Environmental Change*, 7:3, 195–203

Jönsson, Christer, Bo Bjurulf, Ole Elgström, Anders Sannerstedt and Maria Strömvik (1998) 'Negotiations in networks in the European Union', *International Negotiations*, 3:3, 319–344

Kanie, Norichika (2007) 'Middle power leadership in the climate change negotiations: foreign policy of the Netherlands' in Paul Harris (ed) *Europe and global climate change: politics, foreign policy and regional cooperation* (Cheltenham, United Kingdom: Edward Elgar Publishing), 87–112

Kelley, Judith (2004) 'International actors on the domestic scene: membership conditionality and socialization by international institutions', *International Organization*, 58:3, 425–457

Kohler-Koch, Beate (2002) 'European networks and ideas: changing national policies?', *European Integration Online Papers*, (EIoP), 6:6, <http://eiop.or.at/eiop/texte/2002.006a.htm>, accessed 13 February 2008

Krasner, Stephen (1983) 'Structural causes and regime consequences: regimes as intervening variables' in Stephen Krasner (ed) *International regimes* (Ithaca, New York: Cornell University Press), 1–21

Levy, Marc A, Oran R Young and Michael Zürn (1995) 'The study of international regimes', *European Journal of International Relations*, 1:3, 267–330

Lewis, Jeffrey (2005) 'The Janus face of Brussels socialization and everyday decision making in the European Union', *International Organization*, 59:4, 937–971

Lightfoot, Simon and Jon Burchell (2004) 'Green hope or greenwash? The actors of the European Union at the World Summit on Sustainable Development', *Global Environmental Change*, 14:4, 337–344

McCormick, John (2001) *Environmental policy in the European Union* (Basingstoke, United Kingdom: Palgrave)

Michaelowa, Axel (1998) 'Impact of interest groups on EU climate policy', *European Environment*, 8:5, 152–160

Oberthür, Sebastian (1999) 'The EU as an international actor: the protection of the ozone layer', *Journal of Common Market Studies*, 37:4, 641–659

Oberthür, Sebastian (2006) 'The climate change regime: interactions with ICAO, IMO, and the EU burden-sharing agreement' in Sebastian Oberthür and Thomas Gehring (eds) *Institutional interaction in global environmental governance: synergy and conflict among international and EU policies* (Cambridge MA, United States: MIT Press), 53–77

Oberthür, Sebastian and Thomas Gehring (2006) 'Conceptual foundations of institutional interaction' in Sebastian Oberthür and Thomas Gehring (eds) *Institutional interaction in global environmental governance: synergy and conflict among international and EU policies* (Cambridge, Massachusetts: MIT Press), 19–52

Peters, Guy (1994) 'Agenda-setting in the European Community', *Journal of European Public Policy*, 1:1, 9–26

Peterson, John (1995) 'Decision-making in the European Union: towards a framework for analysis', *Journal of European Public Policy*, 2:1, 69–93

Peterson, John (2001) 'The choice for EU theorists: establishing a common framework for analysis', *European Journal of Political Research*, 39:3, 289–318

Peterson, John and Elizabeth Bomberg (1999) *Decision-making in the European Union* (New York: St Martin's Press)

Radaelli, Claudio M (1999) 'The public policy of the European Union: whither politics of expertise?' *Journal of European Public Policy*, 6:5, 757–774

Radaelli, Claudio M (2003) 'The Europeanization of public policy' in Kevin Featherstone and Claudio M Radaelli (eds) *The politics of Europeanization* (Oxford: Oxford University Press), 27–56

Risse-Kappen, Thomas (1995) 'Bringing transnational relations back in: Introduction' in Thomas Risse-Kappen (ed) *Bringing transnational relations back in: non-state actors, domestic structures and international institutions* (Cambridge: Cambridge University Press), 3–33

Scheipers, Sibylle and Daniela Sicurelli (2007) 'Normative power Europe: a credible utopia?', *Journal of Common Market Studies*, 45:2, 435–457

Schreurs, Miranda (2004) 'The climate change divide: the European Union, the United States, and the future of the Kyoto Protocol' in Norman J Vig and Michael Faure (eds) *Green giants? Environmental policies of the United States and the European Union* (Cambridge, Massachusetts: MIT Press), 207–230

Schreurs, Miranda and Yves Tiberghien (2007) 'Multi-level reinforcement: explaining European Union leadership in climate change mitigation', *Global Environmental Politics*, 7:4, 19–46

Sedelmeier, Ulrich (2006) 'Europeanization in new member and candidate states', Living Reviews in European Governance, 1:3, <http://www.livingreviews.org/lreg-2006-3>, accessed 13 February 2008

Sjöstedt, Gunnar (1998) 'The EU negotiates climate change: external performance and internal structural change', *Cooperation and Conflict*, 33:3, 227–256

Skjærseth, Jon Birger (1994) 'The climate policy of the EC: too hot to handle?', *Journal of Common Market Studies*, 32:1, 25–46

Skjærseth, Jon Birger and Jørgen Wettestad (2008) *EU emissions trading: initiation, decision-making and implementation* (Aldershot, United Kingdom: Ashgate)

Tsebelis, George (2002) *Veto players: how political institutions work* (Princeton: Princeton University Press)

Underdal, Arild (2002) 'One question, two answers' in Edward L Miles, Arild Underdal, Steinar Andresen, Jørgen Wettestad, Jon Birger Skjærseth and Elaine M Carlin (eds) *Environmental regime effectiveness: confronting theory with evidence* (Cambridge, Massachusetts: MIT Press), 3–45

Underdal, Arild (2004) 'Methodological challenges in the study of regime effectiveness' in Oran R Young and Arild Underdal (eds) *Regime consequences: methodological challenges and research strategies* (Dordrecht, The Netherlands: Kluwer Academic Publisher), 27–84

Vogler, John (1999) 'The European Union as an actor in international environmental politics', *Environmental Politics*, 8:3, 24–48

Vrolijk, Christiaan (2002) 'A new interpretation of the Kyoto Protocol: outcomes from The Hague, Bonn and Marrakesh', Briefing Paper, Sustainable Development Programme, The Royal Institute of International Affairs, London

Wallace, Helen (1997) 'Introduction' in Helen Wallace and Alasdair Young (eds) *Participation and policy-making in the European Union* (Oxford: Clarendon Press), 1–16

Weale, Albert (1996) 'Environmental rules and rule-making in the European Union', *Journal of European Public Policy*, 3:4, 594–611

Wettestad, Jørgen (2000) 'The complicated development of EU climate policy: lessons learnt' in Joyeeta Gupta and Michael Grubb (eds) *Climate change and European leadership: a sustainable role for Europe?* (Dordrecht, The Netherlands: Kluwer Academic Publishers), 25–45

Wettestad, Jørgen (2005) 'The making of the 2003 EU emissions trading directive: an ultra-quick process due to entrepreneurial proficiency?', *Global Environmental Politics*, 5:1, 1–23

Wynne, Brian (1993) 'Implementation of greenhouse gas reductions in the European Community: institutional and cultural factors', *Global Environmental Change*, 3:1, 101–128

Zito, Anthony R (2001) 'Epistemic communities, collective entrepreneurship and European integration', *Journal of European Public Policy*, 8:4, 585–603

Linking as leverage: emissions trading and the politics of climate change

Richard Benwell

Introduction

On 25 July 1997, the United States (US) Senate unanimously passed the Byrd-Hagel Resolution, which determined that the Kyoto Protocol, then still under negotiation, should not be ratified.[1] As a result, the Protocol itself was never submitted to the Senate for ratification. America was therefore excluded from emissions trading under the Protocol.

On 29 October 2007, 12 US states signed the International Carbon Action Partnership (ICAP) Agreement. ICAP is 'an open forum comprised of public authorities and governments that have established or are actively pursuing carbon markets through mandatory cap and trade systems', designed to promote the best practice in carbon trading and the development of markets across the world.[2] Many US states are now investigating the use of emissions trading markets, most prominently through the Regional Greenhouse Gas Initiative (RGGI) and the Western Climate Initiative (WCI), and many businesses are already participating in emissions trading, notably in the Chicago Climate Exchange (CCX).

Cap-and-trade schemes with an auction-based allocation methodology have been advocated by economists as an efficient environmental policy tool for decades, but it is only recently that policy-makers have acted upon these

---

[1] Senate Resolution 98, 105th Congress, Report 105–154 (1st Session 1997).
[2] <http://www.icapcarbonaction.com/>.

suggestions (Keohane et al 1997, 1). In the 1970s environmental groups scorned the use of emissions trading schemes (ETSs) as a means of preventing climate change, but there has since been a shift of opinion and many emissions trading schemes are now promoted by politicians and accepted by many business and environmental groups across the world (Kerr 2000, 140). It is clear, though, that the decision to adopt environmental targets and then to choose emissions trading as the means of achieving them is not a straightforward step from scientific and economic theory to policy outcome. Rather, a range of political considerations determines which targets will be adopted, the means of achieving them and the timing of the undertaking.

The renewed interest in emissions trading is indicated, first, by developments in the US, such as the emergence of WCI, RGGI and CCX. A second sign is the US delegation's decision at the 13th Conference of the Parties (COP13) to the United Nations Framework Convention on Climate Change (UNFCCC) to join the Bali 'Roadmap' (IISD 2007). A third sign of change is the prominence of climate change issues in the manifestos of the candidates in the US election race in both the Republican Party and the Democratic Party. For example, John McCain has sponsored legislation that promotes a cap-and-trade scheme in the US and campaigns on climate change issues, and Barack Obama advocates 80 per cent cuts in greenhouse gas emissions by 2050 (McCain 2008; Obama 2008).

Many factors have influenced the shift in US domestic political attitudes towards binding caps and international emissions trading. These include the continued advancement of climate science, culminating in the release of the *Fourth Assessment Report* of the Intergovernmental Panel on Climate Change (IPCC) (2007), and progress in the economic assessment of climate change, for example in *The US economic impacts of climate change and the costs of inaction* released last year (CIER 2007). Along with these influences, political pressure has intensified from states, nongovernmental organizations (NGOs) and the public and press for the US government to take action on climate change.[3]

This article, however, examines the influence of a further factor in the change in attitudes towards binding caps and emissions trading: the rapid growth of emissions trading markets themselves, and of the European Union Emissions Trading Scheme (EU ETS) in particular. The preliminary aim is to show how the emergence of well-functioning emissions markets changes the context of future consideration of climate change mitigation.[4] The emergence of multiple, growing carbon markets with increasingly complex infrastructure and sophisticated market players creates a very different setting for both international negotiations

---

[3] For examples of NGO, political and media pressure, see Natural Resources Defense Council (2001); WWF (2001); US Department of Energy 'Scenarios for a Clean Energy Future', ORNL/CON-476 and LBNL-44029 of November 2000; the Climate Stewardship Act of 2003, S.319, 108th Congress (1st session 2003) sponsored by Senators McCain and Lieberman; Climate and Economy Insurance Act of 2005, S. Amendment 868 to H.R. 6, 109th Congress (1st Session 2005), sponsored by Senator Bingaman; and most recently the Climate Security Act of 2008, S.2191/S.3036, 110th Congress (2nd Session 2008), sponsored by Senators Lieberman and Warner.

[4] Under the heading 'emissions trading' I include cap-and-trade under article 17 of the Kyoto Protocol and the EU ETS Directive, the Kyoto project-based baseline-and-credit systems (Joint Implementation and Clean Development Mechanism) and diverse emerging voluntary markets.

and domestic decision-making than the conditions that prevailed a decade ago for the Kyoto talks. Awareness of these changes is creating a domestic constituency for emissions trading in the US and other states, which could influence future acceptance of binding emissions caps and involvement in emissions trading.

The second aim of this article is to demonstrate that ETSs can become more than a tool for the achievement of least-cost emissions reductions. Indeed they can offer considerable added benefits—innovation effects, growth in financial services and marketing opportunities—which can be an incentive for others to participate. Early signs in the EU ETS suggest that the market is becoming a valuable asset in itself, with tremendous potential for growth. The prospect of linkage with EU ETS may now be considered an 'ace' in the negotiating hand of the EU Commission. This advantage could be both passive—the danger of being left out of an emerging global market—or proactive, if the offer of linking is accompanied by sweeteners to the deal. In either case, a more favourable negotiating situation is created.

Emissions trading helps provide a solution to the common resource and public good problems that hamper environmental negotiations. Under normal circumstances, states face the temptation to free-ride in environmental agreements; for instance, the defection of the US in 2001. In order to avoid this, states may turn to tactics such as issue-linkage or sanctions to enforce compliance, but this can be a costly undertaking. Emissions trading, on the other hand, is an inherently 'linked' issue—environment and markets are wrapped up as one. There is an emerging body of evidence that suggests that the emissions trading could provide other benefits, not just cost reduction; in effect, emission trading transforms the issue from a problem of a simple common resource/public good problem to an issue of private property rights and an attractive club good. In other words, apart from the nonexcludable benefits of climate change mitigation, participation in emissions markets may offer certain excludable benefits which can only be enjoyed by members of a 'club' of participating states and businesses. Access to this good can then be used as a bargaining chip in negotiations.

This article does not seek to predict the course of future negotiations on a climate deal, but it does suggest that emissions trading may well be a significant force in future climate mitigation decisions, both as an underlying factor and as a positive bargaining tool, with the prospect of linking. This will require both an historical and a political economy perspective, which takes into account the changing perceptions of environmental regulation and emissions trading over time and which recognizes the interaction of private and public sector actors in the formulation of policy instruments. Structurally, this article begins with a brief exposition of the theoretical role of emissions trading in overcoming the problems of cooperation on climate change mitigation. The next section assesses whether there are indeed added benefits to emissions trading. To test whether these benefits are likely to influence future commitments, I then turn to the question of whether political actors are aware of these changes. Finally, the last section investigates the possibility of using the attractive features of emissions trading proactively to encourage other emitters to adopt climate change targets. I conclude that emissions trading has the potential to influence whether states adopt climate change targets and the nature of those targets. As a factor in the change from 1997 to 2007, this development has so far received little attention.

The goods: common/private, public/club

In order to establish the theoretical basis of how an ETS could overcome barriers to cooperation on climate change mitigation, I set out the nature of the cooperation dilemma. The climate change regime is faced with both a common resources problem and a public goods problem. This section suggests that an ETS can help overcome both challenges by transforming the common good into a private good, and the public good into a club good (see Table 1).

The atmospheric sink is a common pool resource: as a sink for pollution, it has historically been considered nonexcludable. However, it is also finite (leading to rivalry) because there is a limit to the amount of greenhouse gases (GHGs) that can be safely absorbed into the atmosphere before increasing concentrations cause dangerous climate change. The imposition of regulations—the requirement of a permit to pollute and the creation of tradable allowances to govern the level of pollution—can transform this common resource into a private good. These rights can then be purchased by those who value them most.

Climate change *mitigation*, on the other hand, can be described as a public good, because its benefits are nonexcludable and nonrival. By decreasing its exploitation of the atmospheric sink, a state benefits the whole world. This presents the problem that in an agreement to limit emissions, it is tempting for some to free-ride on the mitigation efforts of others. However, the terms of the public good problem posed by climate change mitigation have also been changed by the reality of emissions trading, as participation in trading, and especially in EU ETS, has become a club good. A club good may only be enjoyed by participants. For emissions trading, the public good of climate change mitigation remains nonexcludable. However, the separate benefits of participation in emissions markets can only be enjoyed by members of a scheme (the 'club'), thus helping to counteract the problem of free riding. '[I]ncentives to free-ride on the non-excludable benefits of the global public good can be offset by the incentives to appropriate the excludable benefits of the club good' (Perez 2005, 742). Considering possible links between the World Trade Organization (WTO) and multilateral environmental agreements (MEAs), Perez finds that economic-environmental linkages between regimes are infrequent and limited. In the case of emissions trading, however, this sort of linkage is inherent in the nature of the market mechanism because in order to qualify for participation in trading, certain environmental standards must be met.

The 'club' characteristic of emissions trading is maintained by the rules governing entry into emissions markets. In order to participate in Joint Implementation (JI) or international emissions trading under article 17 of the Kyoto Protocol, it is necessary to be an Annex B Party to the Kyoto Protocol and to

Table 1. The nature of the climate goods with and without emissions trading

	Nonexcludable (without emissions trading)	*Excludable (with emissions trading)*
Rival	Common pool resource (atmospheric sink)	Private property rights (emissions permits and allowances)
Nonrival	Public good (climate change mitigation)	Club good (benefits of emissions markets)

have an 'assigned amount' of allowances to trade. In addition, a Party's Assigned Amount must be properly calculated. It must also have in place a registry and a 'national system for the estimation of anthropogenic emissions by sources and anthropogenic removals by sinks of all greenhouse gases not controlled by the Montreal Protocol', and must be in compliance with a set of reporting requirements.[5] Participation in the Clean Development Mechanism (CDM) does not require Assigned Amount Units (AAUs) to trade (because non-Annex I developing countries do not have Quantified Emission Limitation and Reduction Commitments—QELRCs) but only Parties to the Protocol can take part.[6]

Some participation in the EU ETS is less restrictive. Under article 19 of the EU ETS Directive 'any person may hold allowances' as long as they have an Operator Holding Account or a Person Holding Account, which could theoretically be opened by any individual or group in any of the member states, much like opening a bank account. It is not necessary to be subject to the Emissions Trading Directive (Bossley 2006, 18). There are, however, strict rules concerning the types and volumes of carbon units that can be sold into the EU ETS as laid out in the Emissions Trading Directive and the Linking Directive. Therefore, the full benefits of participation are limited through controls on the ability to sell, rather than through membership or the ability to buy. Most importantly, no allowances can currently enter the EU ETS that were created by non-Parties to the Kyoto Protocol.[7]

The 'club' characteristics of emissions trading, as envisaged in 1997, were largely limited to the benefits of least-cost mitigation. In this sense, participation in emissions trading for compliance with Kyoto would inevitably involve net costs because involvement was contingent on the acceptance of quantified emissions reduction commitments. A notable exception to this is the Porter Hypothesis, which theorized that, over time, environmental protection could beget net commercial advantage as it induced accelerated innovation and hence competitiveness gains (Porter and van der Linde 1995). In a survey of models of cost-benefit of ETSs, however, Oberndorfer and Rennings (2005, 5) found that none incorporated 'Porter-like innovation effects'. In the early years of the climate regime, therefore, emissions trading was factored into cost-benefit analysis only as a tool of cost reduction.

In order to suggest that the consideration of innovation effects and other positive features of ETSs are now influencing interest in participation in emissions trading, three conditions should be satisfied. First, it should be demonstrated that economic costs and benefits are a significant factor in the decision to adopt climate mitigation targets. There are many other considerations that govern state commitments, such as notions of justice and fairness, and concerns over competitiveness. Thus, there is a need to establish that decision-makers have regard to economic effects in their decisions. Second, it should be shown that some positive economic advantages are indeed emerging in existing ETSs. Third, it should be evident that decision-makers are aware of these effects. If it is found that the progress of emissions markets has been successful enough to convince

[5] FCCC/CP/2001/13/Add.2, Marrakesh of 21 January 2002, decision 18/CP.7.
[6] FCCC/CP/2001/13/Add.2, Marrakesh of 21 January 2002, decision 17/CP.7.
[7] Directive 2004/101/EC of 27 October 2004, article 1(2)–(3).

governments and industries of the potential for added benefits from emissions trading—such as those suggested by the Porter hypothesis—then emissions trading could theoretically act as a club good. This could help offset the incentive to free-ride with other considerations as suggested by Perez in the WTO context. The next section turns to the first of these conditions, in order to show how states have factored economic considerations into decisions about their climate goals and the appropriate policy tools.

Emissions trading in cost-benefit analysis

Having discussed the nature of the free-rider problem and the theoretical contribution of an ETS, I now analyse the extent to which that conforms to the practical realities of mitigation action. The first task is to show that economic considerations influence the adoption of ETSs and mitigation targets.

The basic role of cap-and-trade emissions markets is to achieve a certain environmental outcome at the least economic cost. A finite number of tradable allowances are allocated—whether at the state level, the industrial level, or at a lower level—giving polluters for whom it is relatively expensive to reduce their emissions the opportunity to buy allowances from those for whom it is relatively inexpensive. When applied at the substate level, the experience and entrepreneurship available in the private sector can be brought to the problem, opening up more resources and information than governments could access on their own. By lowering an iterative cap over time, emissions are reduced in the most economically efficient manner. Under Kyoto and EU ETS, the cap-and-trade system is complemented by the integration of the project-based flexibility mechanisms. These increase access to low-cost reductions, allow developing countries to begin to experience emissions trading and encourage technological and financial transfers to assist in sustainable development in developing states and transitional economies.

The arguments concerning cost minimization were already brought to the table in 1997, not least by the Americans who had experienced the success of the SO_2 and NO_x programmes implemented in the 1990s by the US Environmental Protection Agency under the Clean Air Act 1990.[8] In the absence of a precedent for an *international* emissions market, though, there could be no guarantee that emissions trading would function as an effective tool. Between 1997 and 2001, there was a range of estimates of the likely costs of Kyoto for the United States. The *potential* cut in the cost of reductions were estimated to be substantial if developed states were able to trade and take advantage of low-hanging fruit in developing nations. Many of the most influential studies, however, included a wide range of estimates that did not include emissions trading, perhaps inflating the apparent extent of the costs involved. Nordhaus and Boyer (2000) suggest that implementation of the Kyoto Protocol without emissions trading would cost US$852 billion. With effective global emissions trading (including the CDM and JI), they claim, the cost could be reduced to US$91 billion (159). Similarly, the Clinton administration predicted that implementation could cost up to 1.5 per cent of GDP without trading or ten times the cost of implementation with

[8] FCCC/TP/2000/2 of 25 November 2000, para 166.

trading (Clinton Administration 1998); the assessment of President Bush's Cabinet-level Review cited estimates that Kyoto could cost 1–2 per cent of GDP by 2010 without trading (Bush Administration 2001). In this way, the question mark that hung over the practical concern about whether international emissions trading could work at all may have made the perception of potential costs of Kyoto much higher in the pre-ETS days.

Even with emissions trading, however, the US considered the 7 per cent reduction against a 1990 baseline mandated by Kyoto to be too much of a threat to the domestic economy and international competitiveness, especially when developing states were not required to accept binding commitments. President Bush argued that 'for America, complying with those mandates would have a negative economic impact, with layoffs of workers and price increases for consumers' and pointed out that 'the world's second-largest emitter of greenhouse gases is China. Yet China was entirely exempted from the requirements of the Kyoto Protocol' (Bush 2001). The prevalent perception among US policy-makers and powerful interest groups in 1997 was that US had relatively little to gain by implementing the Kyoto Protocol, but much to lose. Sunstein estimated that net losses from implementation for the US were perceived to be around US$313 billion (2007, 34). In particular, a large number of significant industrial lobby groups thought that the burden of these losses would fall on key sectors such as energy production and distribution and transport and were able to organize themselves in a highly coordinated lobby against Kyoto (CNN 1997; Boom and Svendsen 2000, 23). Vezirgiannidou (2008) has recently highlighted how these economic concerns were magnified by distributive considerations—the dichotomy between binding QELRCs for developed countries and the lack of substantive targets for developing countries. These considerations were especially influential in the case of the relative gains that would accrue to China, a major economic and political rival, should the US ratify Kyoto.

At the time of the Kyoto negotiations economic and distributional concerns were prohibitive, especially with regard to US participation. Moreover, the untested prospect of international emissions trading could not be factored into calculations with any certainty. Since 2001, however, there have been significant changes.

The next section suggests that the practical successes in emerging emissions markets could add to the benefit side of the equation, offering substantive club good advantages alongside the cost mitigation role. As well as a 'Porter-like' stimulus for green technologies, emissions trading can offer marketing opportunities for firms, 'win-win' efficiency gains, political benefits for governments and new opportunities for financial services, among other benefits, which may not be readily available to nonparticipants. In addition, there is the possibility that those opportunities may be captured by first movers, able to shape the emerging markets to their own preferences. The benefits associated with emissions trading under the compliance regimes have the potential to encourage polluters to make that minimum environmental contribution in order to benefit from participation in emissions trading.

Developments since 1997

This section will explore the benefits that are emerging in connection with emissions trading and illuminate how public and private sector actors in nontrading states are aware of these changes. I suggest that the emergence of functioning international emissions markets since 1997 has made a substantive difference to the way that governments, businesses and other actors think about emissions trading and the reduction of greenhouse gases. I argue that, despite teething difficulties, the experience of real emissions markets has both substantiated the theoretical arguments in favour of emissions trading and uncovered additional benefits that will contribute *another factor* in the many considerations that influence whether a state will adopt environmental targets and emissions trading. The development of global, regional and domestic emissions markets may therefore prove to be a key contextual change in future climate negotiations.

The first and most striking difference in the situation between 1997 and 2008 is that international emissions markets are now a functioning reality. This means that the cost mitigation role of emissions trading can be taken into consideration without the serious doubts that still remained in 2001. A variety of interconnected markets has since emerged. At the highest level, international emissions trading in AAUs under article 17 of the Kyoto Protocol, trade in Certified Emissions Reductions (CERs) as part of the CDM and trade in Emissions Reduction Units (ERUs) under JI have enjoyed growing liquidity. In particular, the CDM saw approximately US$5 billion transacted in 2006 (Capoor and Ambrosi 2007, 3). At the regional level, the EU ETS has emerged as the most active emissions market in the world. At the state level, schemes are appearing or are already active in many states, such as Japan, New Zealand, Australia, Canada, Liechtenstein, Norway and Iceland. Finally, the voluntary emissions market is growing up from the grassroots level, with small volumes transacted so far, but rapid growth. The voluntary retail market reached an estimated 20 Mt in 2006 (Capoor and Ambrosi 2007, 41). Early experiences of trading, such as the BP internal scheme, suggested that ETSs stimulate efficiencies in production and process and Levy has shown how the discourse of environmental-economic 'win-win' situations has been internalized in business discourse (Levy 2003, 93). The strength of these markets suggest that emissions trading is not an ephemeral or small scale phenomenon, but one in which every state and many other actors, such as industry and businesses, will have to participate in future.

The European trading scheme has emerged as the most dynamic market and is being promoted by European politicians as the core of a global carbon market. The creation of a regional cap-and-trade market in the EU by the European Emissions Trading Directive was intended primarily as a means of cost-limitation in the achievement of EU QELRCs.[9] In fact, the EU ETS has grown to be the world's strongest and most liquid emissions market and, despite considerable volatility at times, looks sure to enjoy further rapid growth in years to come. To illustrate, between 2004 and 2005, the value of trade in the EU ETS grew from just under US$8 billion to just under US$25 billion; in comparison, the global value

[9] Directive 2003/87/EC of 13 October 2003.

of emissions markets grew in the same period from around US$10 billion to US$30 billion (Capoor and Ambrosi 2007, 3).

The changes since 1997 go beyond confirming that international emissions trading can serve as a cost mitigation tool. As demonstrated above, participation in mitigation efforts under Kyoto has generally been perceived only as a cost, giving a competitive advantage to nonparties. However, the use of ETSs as the policy tool to achieve mitigation can have a positive economic and competitive effect, through stimulus to innovation, enhanced marketing opportunities and emerging financial markets. With the rapid proliferation of ETSs, the ubiquity of greenhouse gas by-products involved in modern production and consumption patterns suggests that emissions trading may prove to be one of the most important markets in the world. Therefore those regions and states that have opted out of the early experience of emissions trading may experience *negative* distributional effects (from not enjoying the club goods), which will offset *some* of the positive distributional gains that these states initially anticipated by avoiding mitigation targets. The extent of these gains and losses has yet to be quantified and they will not accrue evenly within states, with some winners and some losers. It is becoming clear, however, that important constituencies of support for emissions trading are developing and they are emphasizing the possible gains from emissions trading.

Among businesses, initial fears about the impacts of ET on competitiveness have been allayed by experience and the realization that it is possible to pursue a nuanced approach to emissions trading. In Germany, for example, industrial lobbyists were initially antagonistic toward the adoption of ET, but their position has become more positive (Berg 2008). This change can be attributed to the careful attention to competitiveness by the European Commission, which first employed the blunt instrument of grandfathering allowances (which protected industry, but also limited the usefulness of the scheme), but is now intending to develop a harmonized system of *selective* free-allocation. It has been shown that in the first phases of EU ETS, electricity generators gained substantial windfall profits from the over-allocation of free allowances, which enabled them to pass costs to customers that were essentially nonexistent (Neuhoff et al 2006). The windfall profits in the European power sector were in the order of €5 billion in 2005 and could double that figure in each year of Phase II (Grubb 2007). The reform package proposed by the Commission for Phase III contains a methodology which protects only those industries that are most subject to international competition and cannot pass on costs to consumers.[10]

On the positive side of the equation, there is evidence that the advantages suggested by the Porter Hypothesis (described in the goods section) are in fact appearing and, importantly, that industry *believes* this to be true. As soon as the Kyoto Protocol was ratified, there was a 'Kyoto boost' in renewable energy markets. New Energy Finance (2005) reported at that time that 'the 20 constituent companies of the GEIX [renewable and low-carbon energy technology companies] which are quoted in countries that signed the Kyoto Protocol are up by an average of 21.9%. By contrast, the 30 GEIX constituents quoted in the USA and Australia, which did not sign the Kyoto Protocol, are down by an average of 13.3%'.

---

[10] COM(2008)16 final of 23 January 2008, art 1(8).

In its *Review of EU Emissions Trading Scheme*, carried out by the European Commission Directorate General for the Environment, McKinsey and Ecofys found that half of the surveyed companies said that EU ETS had a strong or medium influence on their innovation decisions (European Commission 2005, 7). This is corroborated by a survey conducted by Point Carbon (2008, i), in which 73 per cent of EU ETS respondents stated that the carbon price was relevant to investment decisions and two-thirds say that EU ETS is resulting in internal emissions abatement.

Significantly, it is not just current participants who are aware of these effects. Some American analysts, for example, are beginning to look enviously at developments in Europe. Daniel Bodansky noted (2002, 2) that 'the U.S. refusal to join Kyoto could disadvantage U.S. firms with technologies to reduce greenhouse gas emissions by limiting their opportunities to develop experience and relations under the Kyoto system'. More recently, The United States Climate Action Partnership has issued a report, *A call for action*, recommending a cap-and-trade ETS on environmental and competitiveness grounds (USCAP 2008).[11] This is a marked change from the situation in 2000, when Boom and Svendsen (2000, 5) found that 'generally, industrial organizations ... argue that global warming is not a serious problem and that taking measures to reduce emissions of GHGs will be very costly to the US', and represents a shift in the lobbying position of major industry in the United States, motivated by a change in their perception of the competitiveness impacts of emissions trading.

The financial services sector is also reporting gains from emissions trading. The report commissioned by the Lehman Brothers on emissions trading and the City of London notes that the UK has benefited from the learning experience of the UK ETS and the EU ETS, saying that 'clear signs of this advantage can be found in the fact that the most actively traded emissions exchange ('ECX') is located and cleared in London, dealing in more than twice the volume of its nearest competitor' (Bossley 2006, 7). This report identifies increasingly sophisticated market features, with the development of spot, term, forward and futures contracts, with options, swaps and spreads, undertaken by increasingly sophisticated players, as legal, technical and financial services move to support the markets. In this way, an emissions trading cluster has developed in London, with smaller centres in Leipzig and in Paris, which may claim a substantial portion of the value of future emissions trading services thanks to their early lead under the EU ETS, ahead of New York and Tokyo, for example.

In addition, emissions trading is beginning to provide considerable opportunities for companies to present themselves as 'green'. The voluntary markets are enabling companies to promote products with 'carbon neutral' or 'zero carbon' labels. Although the benefits of the voluntary markets are not limited in the same way as the compliance markets (and in fact several multinational companies, such as HSBC with operations in non-Kyoto states, are

[11] USCAP includes Alcoa, General Motors Corporation, Caterpillar Incorporated, Ford Motor Company, General Electric, BP America Incorporated, Duke Energy, ConocoPhillips, Pepsi Company, Siemens Corporation, Xerox, Johnson & Johnson, DuPont, Deere & Company, Dow Chemical and other major corporations.

active in the voluntary market), they have tended to centre around the compliance markets both in terms of infrastructure and market activity, with prominent players such as Climate Care and the Carbon Neutral Company basing their operations in the UK. For example, Bodansky (2002, 2) argued that 'the U.S. refusal to join Kyoto could lead to negative public attitudes towards U.S. companies and even consumer boycotts, as well as to trade measures by Kyoto parties against non-parties'.

Linking as leverage

The expansion of emissions markets and the benefits enjoyed by participants may be viewed as contextual, exogenous factors in other states' consideration of QELRCs and emissions trading as well as in future negotiations. This section suggests that there is potential for first movers in emissions trading, especially for the EU as market-maker of the most important emissions trading scheme in the world, to have a more active influence. This active influence could come through the assertion of compatible technical design features in emerging ETSs based on high standards adopted in existing schemes. It may now also be possible for linking deals to be tailored to encourage ambitious objectives through the use of concessions and side-payments. The latter leverage, however, is limited by the problems associated with different kinds of linking, which potentially threaten the integrity of the emissions markets, and therefore both its environmental and economic benefits. This section will focus on the EU ETS, because it is the largest emissions market and because it is possible for the EU to set out with a coherent agenda to influence the way that other states design their own environmental policy tools.

There are many variables in the design of an ETS which will influence its environmental and economic effectiveness. It is possible that trading schemes can be designed in such a way as to give the appearance of action to mitigate climate change for political or marketing purposes, without achieving any environmental improvements. For example, having announced that it is unlikely to meet its Kyoto targets, the Canadian government is keen to appear to be taking some steps towards climate change mitigation and several designs for emissions trading schemes have been proposed. It seems likely, however, that a Canadian scheme will be based on intensity targets and will not have an overall emissions cap (Haites 2007). This means that there will be no certainty of environmental outcome, and net emissions could easily rise even as intensity falls, as it has in the US over the last decade (Sunstein 2007, 41).

It is possible, however, that the adoption of high standards in the largest emissions trading schemes can act as a benchmark and encourage high standards in emerging schemes, because in order for schemes to be linked (and therefore in order for emerging schemes to enjoy the benefits of large schemes) they must have compatible designs or risk prohibitive transaction costs.

The legal basis for linking agreements between the EU ETS and other Annex B countries is contained in article 25(1) of the Emissions Trading Directive, which states that 'agreements should be concluded with third countries listed in Annex B to the Kyoto Protocol which have ratified the Protocol to provide for the mutual recognition of allowances between the Community scheme and other greenhouse gas emissions trading schemes'.[12] This was developed by paragraph 18 of the preamble to the Linking Directive, which creates a duty that

> the Commission should examine whether it could be possible to conclude agreements with countries listed in Annex B to the Kyoto Protocol which have yet to ratify the Protocol, to provide for the recognition of allowances between the Community scheme and mandatory greenhouse gas emissions trading schemes capping absolute emissions established within those countries....[13]

This opened up the possibility of linking EU ETS with developing states and states that are not Parties to the Kyoto Protocol, should they introduce a compatible scheme. Through these linking options, there is an incentive for the designers of any emerging scheme at least to consider compatibility with EU ETS.

The benefits of linking are twofold. Firstly, the opportunity to buy and sell allowances and offsets between schemes creates a broader, more liquid market which can uncover efficiencies and create cost savings for both sides of the agreement. This is recognized in the ETS Directive, which states in paragraph 18 that 'linking the Community scheme to greenhouse gas emission trading schemes in third countries will increase the cost-effectiveness of achieving the Community emission reductions target as laid down in Decision 2002/358/EC on the joint fulfilment of commitments'.[14] Secondly, linking can open access to other benefits of emissions trading, such as those as suggested in the first part of this article.

These benefits, however, can be reduced by inconsistencies in the design of trading schemes and it is this that provides the encouragement for harmonization of design, which can be used to promote the high standards of the EU ETS. For example, it would be possible to link two schemes in which one accepts ERUs for compliance, but the other does not. Either the second scheme would have to accept that the link would create a *de facto* point of entry into the scheme, perhaps undermining the environmental reasoning for not accepting those units in the first place, or there would need to be a 'gateway' that would only allow certain carbon units to pass between the schemes. The first option would involve an environmental cost and the second would involve an economic cost as the creation of a gateway could entail considerable transaction costs (Freestone and Streck 2005, 511).

In order to access the full advantages of the club good of the EU ETS, it would be necessary for other groups to have an emissions trading scheme and to establish a direct link between that scheme and the EU system. This, in turn, would depend on the acceptance of certain standards in order to avoid transaction costs. Some benefits, however, are attainable from partial or indirect linking. The design of emissions markets is crucial to their compatibility for linkage and different sorts of emissions trading will be suitable for different sorts of linkage, with different degrees of access and transaction costs associated. Among the crucial factors for compatibility are:

- *Type of targets*: targets may be expressed in a variety of ways, such as a cap on the volume of emissions (as in the EU ETS) or as intensity targets (such as the

[12] Directive 2003/87/EC of 13 October 2003, art 25(1).
[13] Directive 2004/101/EC of 27 October 2004, para 18.
[14] Directive 2003/87/EC of 13 October 2003, para 18.

proposed Canadian system). Linking schemes with different sorts of target would be technically difficult and could undermine the environmental integrity of the scheme.

- *Price ceilings and floors*: some schemes (such as New Zealand's) have set limits on the price of carbon units, which could cause distortions if linked with a market without price caps and floors.
- *Compliance mechanisms*: a scheme with strict sanctions for noncompliance can be undermined by a more lenient regime. For example, in the EU ETS, noncompliance is met with a fine *and* the obligation to make up the difference in the next period. Linkage to a system with fines alone could act as a buy-out price for emitters.
- *Type of units allowed*: linkage to a scheme that allows the use of carbon units that are not currently allowed will either allow those units into both schemes or require strict limitations on the extent of the linkage.
- *Supplementarity caps*: supplementarity is the principle contained in article 17 of the Kyoto Protocol that most emissions reductions should be achieved domestically, without excessive reliance on trading. Credits from outside linked schemes should be supplemental to internal reductions. If one scheme allows a much greater proportion of external trading, supplementarity could be undermined.

In each of these cases, EU ETS has set a high benchmark:

- the scheme is cap-and-trade (with a fixed target);
- it has no price limitations, there are severe penalties for noncompliance;
- EU ETS does not allow the use of credits for which environmental credentials are unsure, such as CERs from forestry or US Renewable Energy Certificates; and
- the European Commission monitors the observance of supplementarity in the National Allocation Plans of the member states.

This high benchmark, if the EU is willing to defend the integrity of the EU ETS, can act as an incentive to encourage other groups to adopt emissions caps with high standards in order to benefit from the potential to link with the European scheme.

At present, the prospect of linking appears to have had varied influence in the design of emerging trading schemes, many of which seem to have been more influenced by domestic factors. For example, the design of the New Zealand ETS (NZ ETS) announced on 20 September 2007 was powerfully influenced by the special interests of lobbyists from the agricultural and forestry sectors. Credits for afforestation will form a large part of the liquidity of the NZ ETS, but the significant agricultural sector will not be included until 2013 (Ward 2007). So far, no clear timetables for reductions have been set under the scheme. These features may appease domestic interests, but they will hamper the ability of NZ ETS to link directly with EU ETS.

On the other hand, there is some evidence that awareness of the desirability of linking and of the extra costs of linking systems with substantial technical differences seems to be providing a countervailing influence on the design of emerging schemes. As market-maker of the largest scheme in the world, the EU is taking the lead in this process. For example, the International Carbon Action Partnership (ICAP), an open forum intended to steer the development of carbon markets towards greater compatibility, is a Europe-led initiative. Key elements of ICAP's agenda are in tune with EU design preferences, such as the emphasis on cap-and-trade systems, as opposed to offsetting or efficiency-based commitments (ICAP 2007).

The influence of the EU ETS is also apparent in the successful linking operations that have already taken place. Firstly, the terms of the Linking Directive 2004 (which connected EU ETS to JI and CDM) have influenced the evolution of the project-based activities. The pattern of project development has been shaped by the preponderance of EU buyers on the demand side. The EU ETS is linked with the Kyoto flexible mechanisms to allow entry of CERs and ERUs into the scheme, but it has imposed limitations on number and type. This benefits both the suppliers of project credits and the EU member states. European buyers brought demand to the Kyoto markets with 86 per cent of the market share in 2006 and, in return, the use of project mechanisms added liquidity and flexibility for compliance with a supply of competitively priced carbon units (Capoor and Ambrosi 2007, 4). The implications of linking with the EU ETS can be seen in the spread of projects—which have responded to the limitations set by the Linking Directive—that have appeared under the Clean Development Mechanism. Most obviously, article 1(3)(b) of the Linking Directive prohibits the use of CERs and ERUs from land use, land use change and forestry (LULUCF) in the Community scheme.[15] Capoor and Ambrosi (2007, 28) suggest that these EU rules are at least partly responsible for the very small volumes of LULUCF credits have been transacted so far—just one per cent of total volume—and few LULUCF methodologies have been accepted by the CDM Executive Board.

Secondly, on 26 October 2007, Liechtenstein, Norway and Iceland have now concluded a linking agreement with the EU ETS, which will come into effect when the European Economic Area (EEA) states fulfil the requisite technical requirements.[16] This agreement incorporates most of the EU decisions for the functioning of the EU ETS into the EEA Agreement, including the Emissions Trading Directive 2003 and the Linking Directive 2004. Article (2)(l) of the EEA Decision makes any allowances allocated by the European Free Trade Area (EFTA) Surveillance Authority fully fungible in the EU ETS and EU Allowances (EUAs) fully fungible in the EEA. The design of these three trading schemes has been based very closely on the EU ETS and linking has been achieved at a very early stage, showing the influence of the EU lead in the formative stages of the wider European project, which has conformed to the high standards set by the EU ETS.

The EU ETS benefits from a straightforward linking process and the gains of a larger market, whereas the EEA states gain access to that market. There are, however, several important caveats in the acceptance of the Emissions Trading Directive, which can be explained as sweeteners in the deal for the EEA states. For example, article 2(e) of the EEA Decision states the EEA members reserve the right to require payment for more allowances than is permitted under article 10 of the Emissions Trading Directive, while paragraph 11 of the preamble reserves the right for EEA members unilaterally to include carbon capture and storage in their emissions accounting, reducing net emissions for compliance with the emissions

[15] Directive 2004/101/EC of 27 October 2004, article 1(3)(b).
[16] Decision of the EEA Joint Committee No 146/2007 of 26 October 2007.

trading scheme. In addition, Iceland claimed a significant exemption for its fish processing plants under article 2(t), citing Decision 14/CP.7 concerning the impact of single projects and the rules under the EU ETS Directive concerning alternative policies and measures.

This sort of concession can provide an important incentive to parties considering linking with the EU scheme. Fish processing plants are a key industry in Iceland and will be given a degree of protection by their exemption from the scheme at the expense of a small loss to the efficiency of the scheme as a whole. Concession need not take the form of exemptions for particular installations, or the inclusion of carbon capture and storage. Similar sweeteners could be provided by generous EUA allocations, less rigorous monitoring and enforcement requirements or bolder measures such as price guarantees. This sort of trade-off is not unprecedented in the history of emissions trading. Notably, in the negotiations of Kyoto commitments, Russia was offered a high allocation of AAUs and the opportunity to include sinks in its emissions accounting in order to make the prospect of participation palatable (Sunstein 2007, 27).

Of course, just as the side-payments to Russia needed to be carefully balanced against the implications for supply and demand and environmental integrity under Kyoto, so the use of linking and emissions trading concessions as side-payments must be balanced against the implications of such measures for the environmental effectiveness of the EU ETS. The credibility of the scheme and its value as a club are heavily dependent on the robustness of its environmental contribution, and the EU Commission has emphasized that the maintenance of the integrity of EU ETS is a high priority. A communication from the Commission and Council and Parliament concerning the National Allocation Plans for the second phase of EU ETS states that 'a successful EU ETS is of vital importance to sustaining the EU's credibility in relation to the post-2012 climate regime'.[17] The Commission showed itself as willing to act on this principle by requiring substantial changes in most of the Phase 2 National Allocation Plans, including the volume of allowances allocated and the proportion of effort that could be met through the use of flexible mechanisms (Europa 2006). These few examples may be early indications that the EU Commission is willing to wield some of the bargaining power provided by the value of the EU ETS to leverage meaningful commitments from other states in the future, but not at the expense of the integrity of the EU scheme.

Conclusion

On 6 June 2008, although it did not gain enough votes to prevent a filibuster, a majority of senators voted in favour of the Climate Securities Act and John McCain sent a message to Congress stating that 'it appears that for now, the Senate ... will choose to put politics above policy, and Congress will fail to act yet again on this critical issue. But rest assured, we will not give up until we finally succeed in enacting needed, comprehensive cap and trade legislation to address this urgent problem'.[18]

[17] COM(2006)0725 of 29 December 2006, s.3.

[18] Congressional Record—Senate—page s5350, 110th Congress, 6 June 2008. Congressional Record—Senate—page s5334, vote no. 145, 110th Congress, 6 June 2008; the result of the vote was 48 yeas to 36 nays, 60 yeas were required for the Act to progress.

Emissions trading schemes have appeared as a dynamic presence in the economic and environmental calculations of states and businesses in the last decade. There is now real-world evidence of the cost-minimization role of international emissions trading. In addition, there are indications of additional benefits from ET for certain states and sectors. Thanks to limitations on participation in the major schemes—based on the acceptance of Kyoto targets—the chance to benefit from these new opportunities is restricted, making participation in the established schemes an increasingly desirable club good.

Evidence of the potential for established emissions trading schemes to influence the objectives of other states and emerging schemes through leadership and through linking deals is only beginning to emerge. The early indications are mixed, with some schemes developing quite separately from EU ETS, some expressing an interest through initiatives such as ICAP, whereas others are clearly shaped by the intention to link. Nonetheless, whether it is utilized or not, the potential for political leverage does seem to have emerged as part of the first-mover advantage and rapid growth of the European market. This may represent a significant change in the negotiating balance over the last decade, which will complement other advances in public awareness and climate change science, but it is one that has so far received very little attention.

References

Berg, Axel (2008) 'Thinking about transatlantic climate policy', FACET Commentary 9, May, Forum for Atlantic Climate and Energy Talks, Washington DC

Bodansky, Daniel (2002) 'Implications for US companies of Kyoto's entry into force without the United States', Working Paper January, Pew Center on Global Climate Change, Arlington, Virginia

Boom, Jan-Tjeerd and Gert Tinggaard Svendsen (2000) 'The political economy of international emissions trading scheme choice: empirical evidence', Working Paper 00-19, Institute of Economics, University of Copenhagen

Bossley, Liz (2006) *Emissions trading and the City of London* (London: City of London)

Bush, George W (2001) 'President Bush discusses global climate change', Press release, The White House, 11 June, <http://www.whitehouse.gov/news/releases/2001/06/20010611-2.html>, accessed 10 February 2008

Bush Administration (2001) 'Climate change review—initial report', The White House Cabinet-level Climate Change Working Group, <http://www.whitehouse.gov/news/releases/2001/06/climatechange.pdf>, accessed 1 June 2008

Capoor, Karan and Philippe Ambrosi (2007) *State and trends of the carbon market 2007* (Washington DC: World Bank and IETA)

CIER (2007) 'The US economic impacts of climate change and the costs of inaction', October, Center for Integrative Environmental Research (CIER), University of Maryland

Clinton Administration (1998) 'The Kyoto Protocol and the President's policies to address climate change: administration economic analysis', The White House, Executive Office of the President, Washington, DC, July

CNN (1997) 'Negotiators discuss "differentiated" emissions cuts', 2 December

Europa (2006) 'Emissions trading: Commission decides on first set of National Allocation Plans for the 2008–2012 trading period', IP/06/1650, 29 November, <http://europa.eu/rapid/pressReleasesAction.do?reference=IP/06/1650&format=HTML&age>, accessed 10 February 2008

European Commission (2005) 'Review of EU Emissions Trading Scheme: survey highlights', Directorate-General for Environment, Brussels, <http://ec.europa.eu/

environment/climat/emission/pdf/etsreview/highlights.pdf>, accessed 10 August 2008

Freestone, David and Charlotte Streck (eds) (2005) *Legal aspects of implementing the Kyoto Protocol mechanisms: making Kyoto work* (Oxford: Oxford University Press)

Grubb, Michael (2007) 'Lessons and evolution of the EU Emissions Trading Scheme—carbon pricing from theory to practice', Climate Change Policy Seminars, University of Warwick, 8 June, <http://www2.warwick.ac.uk/fac/soc/csgr/activitiesnews/seminars1/climatechangepolicy/>, accessed 10 February 2008

Haites, Erik (2007) 'Linking from the Canadian perspective', Presentation at the Climate Strategies Workshop on Linking, Brussels, 30 October, <http://www.joanneum.at/climate/linking/presentations.html>, accessed 10 February 2008

ICAP (2007) 'Political declaration', 29 October, International Carbon Action Partnership, Lisbon, <http://www.icapcarbonaction.com/declaration.htm>, accessed 10 February 2008

IISD Reporting Services (2007) 'Earth negotiations bulletin', International Institute for Sustainable Development, <http://www.iisd.ca/climate/cop13/>, accessed 10 February 2008

IPCC (2007) *Fourth assessment report: climate change 2007* (Geneva: IPCC Secretariat)

Keohane, Nathaniel, Richard Revesz and Robert Stavins (1997) 'The positive political economy of instrument choice in environmental policy', Discussion Paper 97-25, Resources for the Future, Washington

Kerr, Susan (ed) (2000) *Global emissions trading: key issues for industrialized countries* (Cheltenham, United Kingdom: Edward Elgar)

Levy, David L (2003) 'Business and the evolution of the climate regime: the dynamics of corporate strategies' in David L Levy and Peter J Newell (eds) *The business of global environmental governance* (Cambridge, Massachusetts: MIT Press), 74–104.

McCain, John (2008) 'Stewards of our nation's rich natural heritage', <http://www.johnmccain.com/Informing/Issues/65bd0fbe-737b-4851-a7e7-d9a37cb278db.htm>, accessed 20 February 2008

Natural Resources Defense Council (2001) 'Bush administration errs on Kyoto global warming agreement', <http://www.nrdc.org/globalWarming/akyotoqa.asp>, accessed 20 February 2008

Neuhoff, Karsten, Kim Keats and Misato Sato (2006) 'Allocation, incentives and distortion: the impact of EU ETS emission allowance allocations to the electricity sector', *Climate Policy*, 6:1, 73–92

New Energy Finance (2005) '"Kyoto Effect" boosts European renewable energy company stocks', Press release, 5 April, <http://www.newenergyfinance.com/NEF/HTML/Press/KyotoEffect.pdf>, accessed 10 August 2008

Nordhaus, William D and Joseph Boyer (2000) *Warming the world: economic models of global warming* (Cambridge, Massachusetts: MIT Press)

Obama, Barack (2008) 'The blueprint for change: Obama's plan for America', <http://www.barackobama.com/issues/>, accessed 20 February 2008

Oberndorfer, Ulrich and Klaus Rennings (2007) 'Costs and competitiveness effects of the European Union Emissions Trading Scheme', *European Environment*, 17:1, 1–17

Perez, Oren (2005) 'Multiple regimes, issue linkage, and international cooperation: exploring the role of the World Trade Organization', *University of Pennsylvania Journal of International Economic Law*, 26:4, 735–778

Point Carbon (2008) 'Carbon 2008—post-2012 is now' in Kjetil Røine, Endre Tvinnereim and Henrik Hasselknippe (eds) <http://www.pointcarbon.com/polopoly_fs/1.912721!Carbon_2008_dfgrt.pdf>, accessed 10 August 2008

Porter, Michael E and Claas van der Linde (1995) 'Toward a new conception of the environment-competitiveness relationship', *Journal of Economic Perspectives*, 9:4, 97–118

Sunstein, Cass R (2007) 'Of Montreal and Kyoto: a tale of two protocols', *Harvard Environmental Law Review*, 31:1, 1–65

USCAP (2008) 'A call for action: consensus principles and recommendations from the US Climate Action Partnership: a business and NGO partnership', US Climate Action Partnership, <http://www.us-cap.org/>, accessed 1 June 2008

Ward, Gerri (2007) 'The proposed NZ ETS and the challenges for linking', 30 October, Brussels, <http://www.joanneum.at/climate/linking/presentations.html>, accessed 10 February 2008

WWF (2001) 'New report disproves Bush claims that global warming treaty would hurt US economy', <http://www.commondreams.org/news2001/0712-06.htm>, accessed 20 February 2008

Vezirgiannidou, Sevasti-Eleni (2008) 'The Kyoto Agreement and the pursuit of relative gains', *Environmental Politics*, 17:1, 40–57

Domestic politics and climate change: international public goods in two-level games

Stephan Kroll and Jason F Shogren[1]

Introduction

Environmental threats like carbon emissions do not respect political boundaries. A ton of carbon emitted in Santiago affects climate change the same way as a ton of carbon emitted in Cape Town. Given the global impacts of international public goods, countries should work together: the responsibility to reduce risks to global public goods must ultimately be shared among countries that benefit from those public goods.[2] The challenge raises the classic dilemma: although nations have a common interest to protect themselves, they also have a private incentive to abate pollution as they see fit, which might not match what is best for the globe. For example, the benefits generated by carbon abatement in Japan cannot be denied to anyone anywhere: it is a global public good covering all countries.

According to economic principles, this creates an incentive for one country to free-ride off the abatement effort of others because they cannot be prevented from the benefits, regardless of abatement. The greater the global net

---

[1] We thank Charles F Mason, Tom Crocker, Tommy Stamland, Owen Phillips, Eric Nye, Kevin Siqueira and seminar participants at Clarkson University, and the Public Choice meetings in Charleston for helpful comments on earlier versions. Shogren thanks the Norwegian University of Life Sciences and GREQAM for the financial support. All errors remain ours.

[2] Other global public goods include global justice, trade agreements, international financial stability, cultural heritage, health, knowledge, telecommunications, cyberspace, peace and foreign aid (see for example Kaul et al 1999). See Gilligan (2004) for a discussion of international public good negotiations with more than two countries.

benefits derived from cooperation over pollution control, the more incentive to free-ride, and the more likely that each country under-provides abatement—given that no global police organization exists to enforce an international agreement (see for example Hoel 1991; Carraro and Siniscalco 1993; Barrett 1994).

But this classic free-riding story is missing something. Governments interact with other governments on the international level; they also, however, must address the preferences of their domestic audience (see for example Barrett 1998, 2003). The outcome of an international negotiation over climate protection, for example, is not the end of the game. Governments still need to secure ratification of the agreement within their own countries, and ratification on the domestic level is not a minor outcome when multiple interests exist in a country (see Putnam 1988, the seminal paper on two-level games),[3] as the preferences of the domestic audience can constrain the actions of the international negotiators.[4] A good example is how in 1997 the United States Senate tried to constrain the Clinton Administration during the Kyoto climate change negotiations. The Senate passed the nonbinding Byrd-Hagel resolution 95–0, which stated that the US should accept no climate agreement that did not demand comparable sacrifices of all nations (including India and China). At least since Schelling (1960), bargaining experts have pointed out that these domestic constraints can be helpful for governments in international negotiations. This so-called *Schelling conjecture* implies that domestic constraints give the government of a country involved in international negotiations a higher utility compared to the case without domestic constraints (see Milner and Rosendorff 1997; Pahre and Papayoanou 1997; Pahre 1997).[5] Negotiators can threaten that the deal will fail unless counterparts soften their positions because the home constituency will not ratify the outcome of the negotiation.

This article provides a framework to better understand the nature of these Schelling-domestic threats in international agreements over national contributions to public goods. We use a two-level game theoretic model to pinpoint how domestic constraints affect the outcome of an international bargaining game.[6]

---

[3] Carraro and Siniscalco (1998, 570) and Folmer et al (1998, 21) were among the first to recognize a need for the literature on international environmental agreements to ask about the motivation of acting governments and include the relationship between international agreements and national voters.

[4] Currarini and Tulkens (1997), Siqueira (2003) and Eckert (2003) model transboundary pollution games with domestic audiences. Using a core-theoretic framework, Currarini and Tulkens focus on the stability of an international agreement when the objective function of the negotiator depends on the aggregate pay-off of a winning coalition in power. Siqueira shows a domestic consumption tax on emissions could be inferior if implemented by a government that does not coordinate its policies with other governments. See our comments on Eckert in the text.

[5] Even though this definition is consistent with Milner and Rosendorff (1997), it is a loose interpretation of Schelling's point (1960, 28). A government tries to change the domestic constraints by changing domestic preferences or institutions to improve the bargaining position (Humphreys 2007 is a recent theoretic study of this idea). In Milner and Rosendorff and our model, preferences and institutions are exogenous. For a discussion on when the Schelling conjecture holds, see Tarar (2005) and Hammond and Prins (2006).

[6] Two-level games have been used to analyse international trade (Grossman and Helpman 1995; Milner and Rosendorff 1996, 1997), arms control (Morrow 1991) and macroeconomic policies (Lohmann 1993).

Two-level games are useful because they allow us to define and isolate the crucial economic incentives that affect behaviour. For the non-technical reader, the real world insight gained from using a game theoretic framework comes from the ability to separate *credible* from *incredible* threats. A threat is credible if a country's proposed unilateral action does not decrease payoffs; otherwise, the threat is considered incredible and unlikely to be played. This is the key to our approach and its contribution to substantive knowledge. With a tool to help separate out the credible from the incredible threats, we can logically rule out some possibilities that might otherwise seem like rational choices. Two-level games matter because they can help us understand why governments do not behave as the traditional free-riding model predicts. Now, a seemingly credible action on the international level can be shown to be an incredible threat given domestic constraints. For instance, our results show that if the domestic situation is a *ratification game* (as in the presidential system of the US), contributions do not exceed those of the traditional free-riding game. The government is constrained by the credible threats of its constituency, but is unfettered by the incredible threats. Yet, if the domestic situation is an *election game* (as in the parliamentary system of most continental-European countries), contributions can exceed the free-riding benchmark. Our model shows that the Schelling conjecture only holds when the home government is receiving an offer from the foreign country; if the home country makes the offer, the domestic constraints only work to lower domestic payoffs.

The paper proceeds as follows. The next section defines the scope of our model within the context of the existing technical literature. In Section 3, we first define the traditional one-level benchmark model and then introduce the two-level game with three actors: two governments and one voter. The game on level one is an ultimatum bargaining game: one country makes a proposal, the other chooses to accept or reject. The game on level two is also an ultimatum bargaining game, in which the voter decides whether to accept the proposal.[7] In the next section, we relax the presumption that the home government is a political monopolist in her own country by introducing a candidate from an opposition party, who is in electoral competition with the incumbent. The decision-maker now competes for the votes of a public that watches the outcome of the international bargaining: they compare the retrospective utility from the agreement to the prospective utility that a challenger could have achieved.[8]

Complementary literature

Following Mayer (1992) and others, we model the international negotiations on two dimensions: domestic and foreign contributions to the international public good. The crucial feature of a two-level game is that although the government plays *two* games, it makes only *one* move; this one move determines its payoff in

[7] Reminiscent of the 'monopolistic agenda-setter' literature, the main difference is that the agenda-setter cannot set monopolistic agendas, even though she is a political monopolist in her own country. See for example Romer and Rosenthal (1979), Banks (1990a, 1993) and Morton (1988).

[8] Such a retrospective-prospective voting rule has been used in the political science literature; see for example Bueno de Mesquita et al (1999).

both of the two games.[9] We extend the two-level model by examining a nonzero sum, two-dimensional conflict model on the first level that includes characteristics of both conflict and cooperation.[10] We use the idea that certain outcomes of these negotiations are Pareto improvements for both governments relative to the noncooperative and other outcomes.[11]

This article is closest in spirit to Milner and Rosendorff (1997), and it is worthwhile to note the differences. First, our utility functions are similar to those in the International Environmental Agreement (IEA) literature, whereas Milner and Rosendorff use utility functions from the trade literature with domestic and foreign tariffs. Their optimal solutions are mostly corner solutions since a tariff cannot be below zero. Second, we examine both kinds of international ultimatum bargaining structures, whereas they focus on the one in which the domestic government has the initiative. It turns out that the results from the game in which the foreign government has the initiative differ substantially. Third, our model assumes complete information, whereas Milner and Rosendorff also consider incomplete information, even though they note that 'it is probably true that incomplete information is not overwhelming and that political actors can design mechanisms to obtain almost complete information' (132). The incomplete information case is interesting for the second big question in the two-level game literature: when will agreements not be ratified? In the complete information case this is not an issue, since every proposal is subgame perfect and will always be accepted. Fourth, and maybe most importantly, we include a model with electoral competition, which yields significant differences relative to the basic model. Although the ratification model is important for political decision-making in countries with presidential/majoritarian system and divided government (like the US), in most continental European countries with parliamentary system ratification of, for example, international environmental treaties is often somewhat of a formality in the parliament and nonexistent on the popular level.[12] Here major international decisions are rather subject of election campaigns.

---

[9] Consider briefly the main difference between our model and those within the literature on strategic delegation bargaining games (see for example Fershtman et al 1991; Cai 2000). The bargaining agent does not get any utility directly from the bargaining outcome but only indirectly via a compensation scheme designed by the principal; in contrast, our agent (the government) has its own intrinsic preferences that might differ from those of the principal (the voter). The reason why the principal hires an agent in a bargaining or any other noncooperative game is nevertheless somehow related to Schelling's ideas: the principal can commit himself and show this commitment to the counterpart. See also Segendorff (1998), who uses a level-one game similar to ours.

[10] Iida (1993, 1996), Morrow (1991), Milner and Rosendorff (1996) and Mo (1994, 1995) examine conflict models with one political issue. Putnam (1988) also focuses on these models, noting that things change with the inclusion of a second dimension.

[11] See also Grossman and Helpman (1995), Milner and Rosendorff (1997), Segendorff (1998), Mansfield et al (2000) and Dai (2006).

[12] Popular ratifications, of course, play a role in decision-making in European countries. Recent international agreements on the constitution of the European Union and the common currency have been exposed to national ratifications in several countries, such as the recent Ireland vote. For academic evaluations of this process, see Hug (2004), Pahre (2004) and Hug and Schulz (2005).

The model we develop has some similarities to Eckert's (2003) approach. Eckert examines the strategic advantage a country can enjoy in international environmental negotiations if the government of a region, with relatively high aversion to environmental control, is the representative in these negotiations instead of the federal government and if the foreign government is represented by the federal government.

Finally, this article extends the literature on the domestic economic effects of constitutions. As advanced by Persson and Tabellini, this literature examines how two constitutional features, electoral rules and forms of government, affect economic policy-making.[13] Theory and empirical results imply, for example, that proportional representation and parliamentary regimes lead to more government spending on broad public goods than plurality rules and presidential regimes. Fredriksson and Millimet (2004a, 2004b) apply this theory to environmental policies and find that environmental policies are stricter in countries with proportional representation, and environmental taxes are higher in countries with a parliamentary system (see also Cremer et al 2008, for the effects of political competition within and between political parties on domestic environmental policies).

Benchmark model: Nash equilibrium of the noncooperative game

For comparative purposes, we begin by defining the classic one-level benchmark model of international public good provision. Two countries, Home and Foreign, choose contributions A_i, i = H, F, to an international public good, such as the abatement of CO_2 emissions, where the contributions are the national abatement measures (see for example Hoel 1991). Following Milner and Rosendorff (1997), assume the governments make choices either in an ultimatum bargaining game (one offers, the other accepts or rejects), or independently if negotiations break down. Also assume that for now neither government faces a domestic constraint; each country acts as a unitary rational actor.

Each government derives benefits $B_i(A_H + A_F)$ from the total level of the public good, and incurs costs $C_i(A_i)$ from the economic consequences of its contribution.[14] Assume benefits are weakly concave and increasing ($B_i' > 0$, $B_i'' \leq 0$), and costs are strictly convex and increasing ($C_i' > 0$, $C_i'' > 0$). Without loss of generality, assume $B_H = B_F = B$. The payoff function for government i is

$$U_i(A_H, A_F) = B(A_H + A_F) - C_i(A_i). \tag{1}$$

Consider first the noncooperative game between the governments. The Nash equilibrium (A_H^N, A_F^N) is the pair of best contribution levels for each government, holding the other's contribution level fixed. Each government i chooses its Nash strategy by solving $\max_{A_i} U_i(A_H, A_F)$, $i \in \{H, F\}$, yielding the first-order conditions $B' = C_i'$.[15]

---

[13] See for example Lizzeri and Persico (2001), Milesi-Ferretti et al (2002), Persson (2004), Persson and Tabellini (1999, 2003, 2004a, 2004b) and Persson et al (2000).

[14] The assumption that the aggregation of the international public good follows a summation technology can matter. The affect of domestic constraints on the supply of public goods when the aggregation is based on a different technology (for an example of a best-shot or a weakest-link scenario see Sandler 1997, chapter 2) remains an open question.

[15] The 'ideal point' of each country is given by its own Nash equilibrium contribution and by an infinite contribution of the other country.

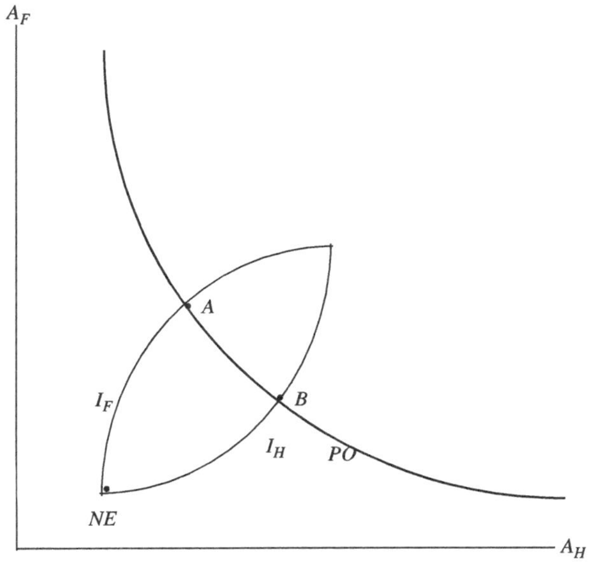

Figure 1. The benchmark model

Pareto improvements for the two governments are possible. Each point in the set PO of Pareto optimal points for H and F, $\{(A_H^{PO}(\lambda), A_F^{PO}(\lambda)), \lambda \in (0,1)\}$, solves $\max_{A_H,A_F}(\lambda U_H + (1-\lambda)U_F)$. This leads to the first-order conditions $B' = \lambda C_H{}'$ and $B' = (1-\lambda)C_F{}'$. Given the assumptions on B and C_i, the contribution levels yielded from these first-order conditions must exceed those in the Nash equilibrium, which is not part of the set of Pareto optimal points. Consequently, there exists an interval $[\underline{\lambda}, \bar{\lambda}] \subseteq [0,1]$ such that for each $\lambda \in [\underline{\lambda}, \bar{\lambda}]$ both governments are better off than at the Nash equilibrium. The limits of this interval are determined by the indifference curves, on which each point makes a government as well off as the Nash equilibrium.

Figure 1 shows the Nash equilibrium (NE) and the indifference curves for the home government, I_H, and the foreign government, I_F. All Pareto optimal combinations for the two governments are elements of the set PO, but generally only the points between the indifference curves are Pareto improvements over the Nash equilibrium, and on PO only the combinations between points A and B are preferred over NE by both governments.

Now consider an ultimatum bargaining game without domestic constraints. Let government i make the offer, and let government j choose to accept or reject this offer. The maximization problem of government i is:

$$\max_{A_H,A_F} U_i(A_H, A_F) \qquad (2)$$

$$\text{s.t. } U_j(A_H, A_F) \geq U_j(A_H^N, A_F^N)$$

The constraint binds because otherwise government i proposes $(A_i, A_j) = (A_i^N, \infty)$, which makes government j worse off relative to the Nash equilibrium. Point A (B) in Figure 1 illustrates the solution to (2) when the home (foreign) government makes the proposal.

The ultimatum game with domestic constraints

Now assume an additional actor enters this game, a median voter or median legislator in the home country, who has the power to veto the agreement negotiated on the international level.[16] This model assumes the home government acts domestically as a Romer-Rosenthal-Banks agenda-setter. She worries about whether (a) her policy, the agreement negotiated with the foreign government, is going to be ratified by the voters/legislators, or (b) she has to install the default policy instead, the Nash-equilibrium of the noncooperative game.[17] This structure captures a presidential system with a divided government (for example the US), or a country that uses referendums for important political decisions (for example Chile).

The median voter has a utility function similar to (1):

$$U_V(A_H, A_F) = B(A_H + A_F) - \gamma C_H(A_H), \tag{3}$$

with γ as a cost parameter that can differ from 1. Depending on the interpretation of the game, several explanations exist for the difference between the executive's and the median voter's utility functions. If we interpret the additional actor as a median legislator then the difference could stem from a divided government: the executive belongs to a different party than the median legislator and therefore views the cost of a contribution to the international public good differently (Milner 1997, 37–42). If the domestic game is a referendum game and the third actor is a 'median citizen', then different interpretations are possible. For example, the executive (or her advisors) could have better knowledge of the good in question (Portney 1992; Pollak 1998) or the executive could be captured by special-interest groups that do not represent the median voter's interests.

The proposing government i now solves the maximization problem:

$$\begin{aligned} &\max_{A_H, A_F} U_i(A_H, A_F) \quad \text{s.t.} \quad U_j(A_H, A_F) \geq U_j\left(A_H^N, A_F^N\right), \\ &U_V(A_H, A_F) \geq U_V\left(A_H^N, A_F^N\right), \end{aligned} \tag{4}$$

where $i, j \in \{H,F\}$, $i \neq j$. The other government, which wants to get at least the same utility as in the Nash equilibrium, imposes the first constraint; the second constraint stems from the voter with the same reasoning.

Consider first the ultimatum game with the home country, H, as proposer.[18] One can show the Schelling conjecture does not hold in this case—H's maximization problem includes an additional constraint, which is summarized below.

Proposition 1: In the ultimatum game with domestic constraints and with the home government as proposer, the home government will either

[16] In the following we will use the feminine pronoun for the Home government H, the male pronoun for the median voter and the neuter pronoun for the foreign government F. See Tarar (2001) for a discussion of the case if F faces a domestic constraint as well.

[17] This is a key assumption. See for example Mansfield et al (2000) who assume the median legislator plays the noncooperative game with the foreign government.

[18] This case is similar to the tariff-setting game in Milner and Rosendorff (1997), with the difference that in their game F and V play the noncooperative game instead of F and H.

be equally well or worse off compared to the case without domestic constraints.

Proof: Since we know already that at least one constraint in (4) is binding, there are three different cases to distinguish:

(a) The first constraint is binding, the second constraint is not. H will propose the combination that maximizes the optimization problem (2), F accepts and V does not reject the agreement—there is no difference to the case without domestic constraints.
(b) The first constraint is not binding, the second constraint is binding. This case is impossible. Increasing A_F by ϵ from such a point (by holding A_H constant) makes F worse off (which is possible, since F's constraint was not binding) and H and V better off. Therefore, this point is not a solution to the constrained maximization problem (4).
(c) Both constraints bind. H can only be either equally well or worse off.

Proposition 1 is stronger than Milner and Rosendorff's (1997) Proposition 3, which also uses an ultimatum game with domestic constraints and H as proposing government. They write that the 'Schelling conjecture holds only when [H and V] have fairly similar preferences'. But our Proposition 1 implies that the Schelling conjecture never holds in such a game. The reason for this contradiction is that Milner and Rosendorff, when discussing the Schelling conjecture, drop the assumption that the Home government makes the proposal and the Foreign government can only react by accepting or rejecting the proposal.[19] Their model would have produced the same result if they had stayed with the ultimatum game throughout.[20]

Less obvious is the question of whether the levels of contribution have to be equal, lower or actually higher relative to the benchmark game without domestic constraints. Proposition 2 summarizes the answer.

Proposition 2: The overall level of contributions to the public good in the ultimatum game with domestic constraints, are less than or equal to contributions in the game without domestic constraints.

Proof: When only the constraint imposed by the foreign government binds, the contributions are identical to the case without domestic constraints. The only interesting case is when both constraints are binding. Here F's and V's relevant indifference curves intersect at the Nash equilibrium and the constrained optimum, at which the contributions are greater than the Nash equilibrium. Therefore, one needs to only show that F's curve is concave everywhere, and V's curve is convex everywhere, which is accomplished by taking the derivatives of the slopes of the indifference curves and signing them.

---

19 This becomes clear when one looks at their Figure 1, in which they compare a point *x* on the contract curve with a point *b* that is the constrained optimum: in their Figure 1 H would prefer *b* over *x* but *x* cannot be the result from an unconstrained optimization á la (2) to begin with.

20 Mo (1994) finds that domestic constraints are a bargaining advantage if the domestic constituents are neither too weak nor too strong.

Figure 2 also includes a possible indifference curve for the median voter, I_V. Based on how we draw the curve here, it does not affect the outcome of the international ultimatum game. But with a larger γ (and therefore a binding constraint), combination A is now unfeasible, and the home government must propose combination C in Figure 3. In summary, domestic constraints do not make either government better off, the Schelling conjecture does not hold, and the contribution levels cannot be higher.

While Milner and Rosendorff (1997) also consider the ultimatum game with H as proposer, they do not examine the case in which F makes the proposal, and in which H and V can only accept or reject this proposal. The results from this game, however, do not have to be equivalent to the game we just consider. As we show below, the results regarding the validity of the Schelling conjecture and the contribution levels differ substantially. This result occurs because in the ultimatum game with F as proposer, the constraints are now 'in the same direction'—H and V both prefer higher contributions by F, *ceteris paribus*, and lower contributions by H (up to a certain level). Let F solve the maximization problem:

$$\max_{A_H,A_F} U_F(A_H,A_F) \quad \text{s.t. } U_H(A_H,A_F) \geq B\left(A_N^N,A_F^N\right) - C_H\left(A_H^N\right),$$
$$U_V(A_H,A_F) \geq B\left(A_N^N,A_F^N\right) - \gamma C_H\left(A_H^N\right) \tag{5}$$

Now, at least one constraint has to be binding, since otherwise F proposes$(A_H,A_F) = (\infty, A_F^N)$. Can both constraints bind simultaneously? Yes, but only if $\gamma = 1$. To see this, transform both constraints into equalities and rearrange:

$$B(A_H,A_F) - B\left(A_H^N,A_F^N\right) = C_H(A_H) - C_H\left(A_H^N\right) \tag{6a}$$

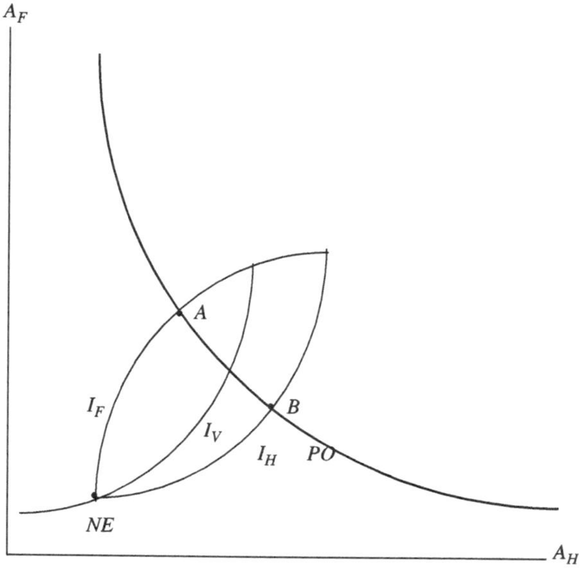

Figure 2. The ultimatum game with domestic constraints I

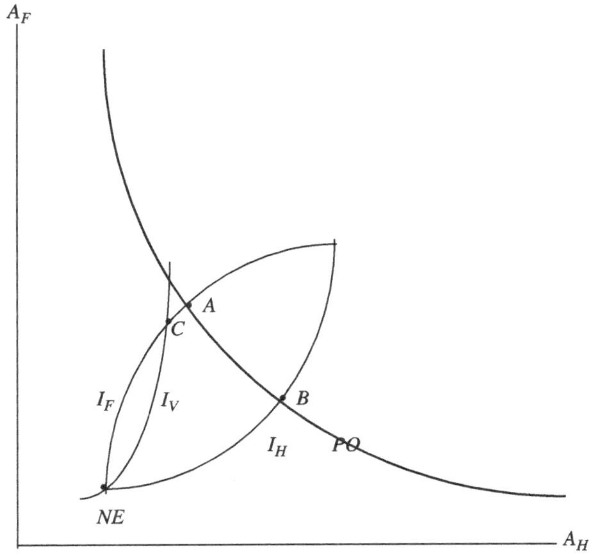

Figure 3. The ultimatum game with domestic constraints II

$$B(A_H, A_F) - B\left(A_H^N, A_F^N\right) = \gamma C_H(A_H) - \gamma C_H\left(A_H^N\right). \tag{6b}$$

If $\gamma = 1$, the two constraints coincide. If $\gamma < 1$, the first constraint is stronger and therefore binding (F proposes combination B in Figure 2); if $\gamma > 1$, the second constraint is stronger and therefore binding (F cannot propose B anymore).

This observation has an important effect on the Schelling conjecture and the levels of contribution summarized below:

Proposition 3: In the ultimatum game with the foreign government as proposer and with domestic constraints, the overall level of contributions is smaller and the home government is better off compared to the same game without domestic constraints, if the domestic constraints are stronger than the constraints given by the home government; thus the Schelling conjecture holds.

Proof: See Appendix.

The basic idea behind Proposition 3 is that F is implicitly bargaining with V even though he is explicitly bargaining with H (as long as V is more 'hawkish' and has a higher cost parameter than H). He knows everything V accepts H will accept too since H will always be better off due to V's strong constraint.[21] This is in contrast to other models in the literature where V's constraint might actually make the home government worse off, since it moves the outcome further away from the home government's preferred point. In our model with contributions to a public

[21] Mansfield et al (2000) avoid this by stating in their footnote 9 that 'we require that the international negotiators choose a ratifiable offer that lies on [PO] when such an agreement exists'. See the discussion in Dai (2002) and Mansfield et al (2002).

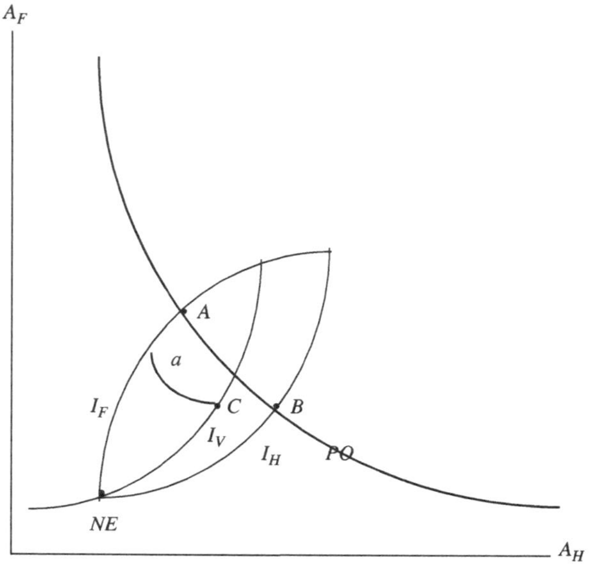

Figure 4. The ultimatum game with domestic constraints III

good, however, the government does not have a (reasonable) preferred point, and the home government always prefers the other country to make higher contributions to the public good, even though it does not prefer it as strongly as V.

Figure 4 shows the case with a $\gamma > 1$. Curve *a* shows the combinations at which F's and V's indifference curves are tangent to each other for this particular γ. F proposes point *C*.

Proposition 3 raises an additional question. Since the home government gains from having a legislature or median voter that is more 'hawkish' than herself (for instance the Schelling conjecture holds), does she gain even more the greater the difference between her cost parameter and the median voter's? Proposition 4 suggests that 'it depends'.

Proposition 4: If the cost parameter of the median legislator or median voter, γ, is greater than 1, an increase in γ, (for example through an election) does not necessarily increase the utility of the home government.

Proof: If $\gamma = 1$, H gets the same utility as in the Nash equilibrium. When γ approaches ∞, H's utility also approaches the benefits she gets from the Nash equilibrium. To see this, note that for $\gamma \rightarrow \infty$ equation (6b), which is a transformation of the binding constraint for maximization problem (5), can only be fulfilled when $(A_H, A_F) = (A_H^N, A_F^N)$. Since the indirect utility function for H is (a) continuous for all γ, (b) for $\gamma \rightarrow \infty$ equal to the utility H receives from the Nash equilibrium ($\gamma = 1$), and (c) larger at some point in between $\gamma = 1$ and $\gamma \rightarrow \infty$ (the Schelling conjecture), this indirect utility function must have a maximum for a certain $\gamma = \tilde{\gamma}$ and decline in γ for $\gamma > \tilde{\gamma}$.

Proposition 4 states that even though the Schelling conjecture holds for all $1 < \gamma < \infty$, cost parameters of the median legislator or median voter exist such that H wishes the constraint imposed by V were not as strong. The executive does not like it when she and the median legislator have become too different.

We have explored some reasons why in a system with a divided government the contributions to an international public good can sometimes be lower than what the executive of one country wishes them to be. Contributions can never be higher compared to the case in which there are no domestic constraints. Higher contributions can be explained, however, if the game on the domestic level is different, which we now consider.

The ultimatum game with domestic electoral competition

We now consider a parliamentary system as a domestic institution. The executive and the median legislator are of the same party or at least the same party coalition, and a ratification by the parliament is sometimes merely a formality due to party discipline.[22] Now the executive is more concerned with how voters perceive the agreement and how they will react in the next general election rather than ratification. Voters cannot change a ratified agreement, but they can punish the government by replacing it with an opposition party or politician. The government should account for these electoral consequences when she bargains on level one.

Political behaviour in Germany during the run-up to Kyoto protocol negotiations is one example of how upcoming elections can affect a government's agenda for an international conference. The conservative German government pushed for 'greener' outcomes at the Kyoto conference. One could interpret their choices as being based on the belief that the opposition parties (including the Greens) could topple the government in the elections eight months after Kyoto, which they did. Such behaviour can be captured by an electoral competition on level two, not by a ratification game. As we show below, this is a case in which a set of cost parameters can induce the home government to propose an agreement that calls for higher abatement levels than what it would have proposed without electoral competition.

Now consider a simple form of electoral competition. In addition to the three actors from the ratification model—the two governments and the median voter (who is here now really a voter and not a legislator)—include another actor in this game, an opposition party or politician. Assume the voter is retrospective in his voting decision, and that he only cares about one issue when he votes—the terms of the agreement on the contributions to the international public good. When the two governments cannot come up with an agreement, however, his voting decision will be independent of the outcome of the noncooperative game. The outcome of the noncooperative game can be thought of as the status quo that home and foreign government had played before the negotiations started and that they do not change once the negotiations have broken down.

[22] For a discussion on how far the 'divided government' discussion also concerns parliamentary democracies see chapter 10 of Alesina and Rosenthal (1995), Milner (1997, 37–42) and Pahre (1997). See also the recent work on the effects of parliamentary versus presidential government forms on domestic public good contributions cited in footnote 13.

When an agreement is reached, the voter compares the utility from this agreement to what he thinks he would have received had the opposition been in power, and he votes accordingly. The opposition herself is passive and does not make any decisions on her own. The voter only has a perception of her cost parameter and can calculate a guess of what she would have done if she had been in negotiations with the foreign government. This unadorned picture of electoral competition is useful because it incorporates the basic point that the home government's bargaining behaviour depends on how the voter sees the opposition.[23]

Again consider an ultimatum bargaining game between the two governments on level one. The new player—the domestic opposition O—has a utility function similar to (3):

$$U_V(A_H, A_F) = B(A_H + A_F) - \delta C_H(A_H). \tag{9}$$

The sequence of moves in this game is:

- *Stage 0*: V conjectures δ, which is common knowledge.
- *Stage 1*: H proposes a combination of contributions, (A_H^{HE}, A_F^{HE}).
- *Stage 2*: F either rejects or accepts this proposal. If F rejects it, H and F keep the status quo, (A_H^N, A_F^N), and the game ends. If F accepts the proposal, the proposal becomes the agreement, and the game moves to stage 3.
- *Stage 3*: V observes the agreement and either votes for H or O.

In Stage 0, we assume it does not matter for the model how V arrived at his conjecture and whether this conjecture is correct,[24] only that V uses δ to calculate what O would have proposed to F if she had been in power, (A_H^O, A_F^O), and that H knows how V perceives δ. In stage 3, V calculates $U_V(A_H^{HE}, A_F^{HE})$, and compares this with the utility from the conjectured agreement, $U_V(A_H^O, A_F^O)$. If $U_V(A_H^{HE}, A_F^{HE}) \geq U_V(A_H^O, A_F^O)$, then V votes for H; if $U_V(A_H^{HE}, A_F^{HE}) < U_V(A_H^O, A_F^O)$, then V votes for O.[25]
According to these assumptions, the voter can, after learning δ, calculate (A_H^O, A_F^O) similar to (2):

$$\max_{A_H, A_F} U_O(A_H, A_F) \tag{10}$$

$$\text{s.t. } U_F(A_H, A_F) \geq U_F\left(A_H^N, A_F^N\right).$$

[23] Two other papers model two-level games with an explicit electoral stage: Morrow (1991) also assumes that the voter has a certain cut-off utility for the case a treaty (on arms control in his paper) between the two countries has been signed and a different cut-off utility for the case such a treaty has not been signed. These cut-off utilities, however, are independent from any electoral considerations on the voter's part. Governments and voters in Smith and Hayes (1997) have preferred points and indifference circles around them: if one government gains, the other loses, which is not so in our public good environment.

[24] Stage 0 could be a signalling game in an electoral campaign. See for example Banks (1990b) or Harrington (1992).

[25] The assumption that voters use such a retrospective-prospective voting rule can be traced back to Downs (1957). See also Williams (1994) and Bueno de Mesquita et al (1999). Also note that there is an implicit assumption in this model that a voter does not 'punish' a government if he or she fails to come up with an agreement. This assumption seems reasonable in cases where the nonagreement would not get a lot of media coverage.

As before, the constraint is binding and O would propose a combination on F's indifference curve that goes through the Nash equilibrium. Exactly where this point is depends on δ. It follows from (8) that both contribution levels are smaller the larger δ is. If $\delta = 1$, O would propose the same point that H proposes without constraints. If $\delta > 1$, O proposes a combination with smaller contribution levels; if $\delta < 1$, O proposes a combination with larger contribution levels.

Now recall the assumption made about V's voting decision: he votes for H if he gets a higher utility from H's agreement with F than what he thought he would have received from an agreement O had struck with F; otherwise he votes for O. H has an advantage as incumbent that O does not have—H can react to V's voting plans with deeds instead of talk. Therefore, H is now facing the following problem:

$$\max_{A_H, A_F} U_H(A_H, A_F) \quad \text{s.t.} \quad U_F(A_H, A_F) \geq U_F\left(A_H^N, A_F^N\right), \qquad (11)$$
$$U_V(A_H, A_F) \geq U_V\left(A_H^O, A_F^O\right),$$

where the argument in $U_V\ (A_H^O, A_F^O)$ is the voter's utility from the solution to O's maximization problem (10).[26] H proposes a combination of contributions that makes V indifferent between voting for her and voting for O. What this combination is depends on the parameters δ and γ.

Note V has a preferred point on F's indifference curve that goes through the Nash equilibrium, a point that gives him the highest utility. This point depends on γ: the larger γ is, the smaller the contributions in the combination V prefers. If $\gamma = 1$, V's preferred point on F's indifference curve is exactly the one that H would have proposed without domestic constraints.

We can now distinguish between different cases regarding V's and O's cost parameters and their impact on H's proposed agreement:

- $\gamma > 1 > \delta$ or $\delta > 1 > \gamma$. In either case, the home government's preferences are closer to the voter's preferences than the opposition's preferences are conjectured to be. The agreement H would propose without domestic constraints is closer to V's preferred point on F's indifference curve than O's preferred point is, and therefore the second constraint in (11) is not binding.
- $\gamma > \delta > 1$ or $1 > \delta > \gamma$. In either case, the opposition's preferences are closer to the voter's preferences than the government's preferences are, and the government has to account for this distance. Now V's preferred point on F's indifference curve is closer to O's preferred point than H's. Since V's utility decreases with distance to its preferred point on F's indifference curve, H has to propose O's preferred point to make V indifferent between H and O.
- $1 > \gamma > \delta$ or $\delta > \gamma > 1$. Now V's preferred position on F's indifference curve is sandwiched between O's and H's preferred positions. Only if V's bliss point is closer to H's than to O's, can H propose the combination she would have

[26] Including the additional constraint that V has to be better off compared to the Nash equilibrium does not change the main results. Note also that the second constraint in (11) means the political topic the countries are bargaining over is not important enough for H to risk losing elections. One could imagine that H is only partially willing to give up his convictions. This could be modelled by adding a positive constant to the left-hand side of the second inequality, and the results on the following pages, while fundamentally still intact, would be weaker with a larger constant.

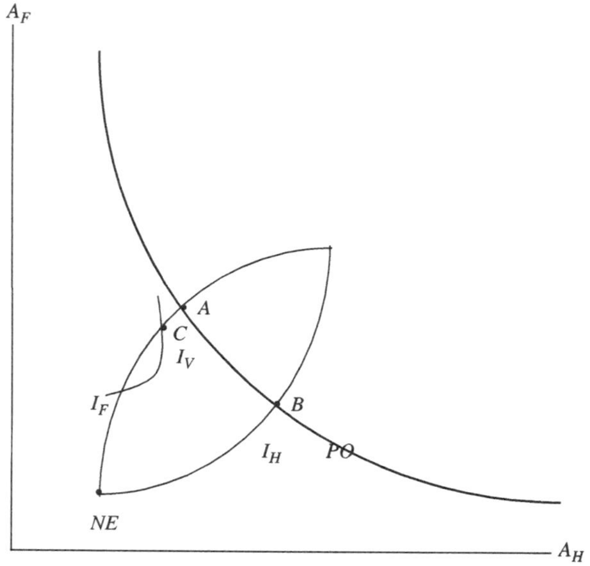

Figure 5. The ultimatum game with elections I

proposed without domestic constraints.[27] But there exist $\bar{\gamma}$ and $\underline{\gamma}$ so that for $1 > \gamma > \bar{\gamma} > \delta$ and for $\delta > \gamma > \underline{\gamma} > 1$, V's bliss point is closer to O's preferred combination. For those parameter values, H has to propose a point closer to V's, so that V is indifferent between voting for O and for H.

In some of these cases we see a result one cannot observe in the pure ratification model. The home government must propose higher contributions to the international good relative to those proposed without domestic constraints, summarized below.

Proposition 5: If the ranking of V's and O's cost parameter is $\gamma > \delta > 1$ or $\delta > \gamma > \underline{\gamma} > 1$, where $\underline{\gamma}$ is defined as above, then H has to propose a combination of contributions to the international public good, in which the contribution levels of both countries are higher than in the baseline case without domestic constraints at all. The Schelling conjecture does not hold in this case; H is worse off compared to the baseline case.

Proof: The only step left to prove is to show that the utility of each domestic actor is single-peaked along F's indifference curve, which means to prove that the utility of each domestic actor is declining the further a combination of contributions on F's indifference curve is away from his or her preferred point. But this follows automatically from the shapes of the indifference curves—all domestic indifference curves are convex while the foreign indifference curve is concave as shown in the proof to Proposition 2.

---

[27] In all these three cases, 'closer' does not refer to a Euclidean distance but to a difference in terms of utility.

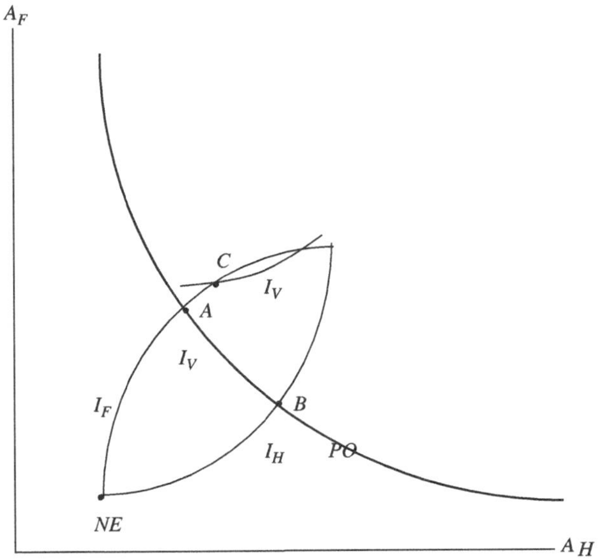

Figure 6. The ultimatum game with elections II

Figure 5 shows the case $1 > \delta > \gamma$ or $1 > \gamma > \bar{\gamma} > \delta$, Figure 6 illustrates $\gamma > \delta > 1$ or $\delta > \gamma > \underline{\gamma} > 1$. In all cases, H proposes C. One interpretation of Proposition 5 is for the case in which the public good is international pollution control and the contributions are abatement measures of the individual countries. Given certain parameters, a government uninterested in high pollution abatement could be forced to propose relatively high abatement levels, if the domestic median voter is 'greener' than the domestic government, and if this voter also perceives the opposition to be 'greener.' Again this case fits Germany and Kyoto: the conservative government under Helmut Kohl did not appear to be fond of 'green' policies but was a prime force behind a 'green' European stance at the Kyoto negotiations.[28] This is consistent with Tjernstöm and Tietenberg (2008) who report on attitudes towards climate change, which we capture by γ.

Conclusion

This article has examined the influence of different domestic constraints on the levels of contributions to an international public good and on the utility a government receives from the outcome of the international negotiations. Our two-level model shows how the structure of the domestic game—ratification or election—can impact the level of contributions. We show that a ratification game at the domestic level can prevent contributions from being greater than what they would have been in the standard noncooperative model without domestic constraints. We also demonstrate that in an election game, in which an opposition and a median voter have smaller costs from the contribution of a country,

---

[28] The empirical study by Jensen and Spoon (2007, 1) finds policy preferences of governing parties, in particular green parties when in the government, 'are significant predictors of [domestic] progress towards GHG [greenhouse gas] emission targets'.

contributions can exceed those from the baseline model. We further show that the Schelling conjecture does not always hold: a government does not necessarily gain from having domestic constraints. The conjecture fails when the government is in a strong position internationally, since the domestic constraint is exactly that—an additional constraint to Home country's maximization problem. If the government is in a weak position, however, the constraint imposed by a more 'hawkish' audience makes the executive better off.

In line with many other articles on two-level games (for example, Milner and Rosendorff 1997, or Tarar 2005), we use an ultimatum game (called 'Take-it-or-leave-it' [TILI] game in most of the early two-level game papers) to describe the interaction between the governments. Naturally this is only a crude picture of international interactions, used for parsimony, since the qualitative character of the main results remains for more complex negotiation games on level one. Our model uses the two most extreme forms, where one 'strong' country makes the offer and the 'weak' country can only react. In more elaborate models, the power structure would be less one-sided. The outcomes in the ratification games of such models would be on a contract curve between the outcomes in our extreme cases, points A in Figure 2 and C in Figure 3 on the one hand, and point C in Figure 4 on the other hand (assuming that the domestic constraint is binding; if it is not binding then all outcomes are of course between A and B on the curve *PO*). The contribution levels would therefore be below the corresponding contribution levels if there had been no domestic constraint, and the main results transfer; if ratification has any effect, then it can only 'pull' contributions down.

While we do not provide statistical analysis to support our claims, we contend the results have a bearing on empirical analyses of international public good negotiations and agreements, particularly for environmental issues such as global climate change. The post-Kyoto process is still in flux, and it is imperative to understand what has driven the negotiations so far. Our results are consistent with behaviour of the American negotiators who were constrained by legislative ratification (see the 95–0 vote in favour of the Byrd-Hagel resolution) and negotiators for European governments that were influenced by electoral concerns (as Anderson 1999, 219, notes, '[v]igorous green parties are nipping at the heels of their governments'). Of course, alternative explanations could also be provided for the different behaviour of countries. Recent research has pointed to differences in, for example, attitudes (Tjernström and Tietenberg 2008), availability of criteria to judge risks (Sunstein 2006), benefit-cost ratios of climate change policies (Nordhaus and Boyer 2003) or the impact of interest groups (Markussen and Svendsen 2005). Some explanations compete with and some complement each other.

An empirical analysis to test our theory should follow the research line of Persson and his coauthors on domestic policies and use the electoral system as a variable to test for their impact on international policies. For example, Persson and Tabellini (1999) find theoretically and empirically that countries with a majoritarian electoral system and presidential regime have a smaller level of (domestic) public goods and government size than their counterparts in countries with proportional electoral system and parliamentary regime. For a precise test of our propositions, other variables in addition to the electoral system would have to be included as well. One will need to include instrumental variables for the different cost parameters in our model; the lower contribution levels in the ratification case and the higher contribution levels in the election case only hold under a certain combination of these parameters. A combination of two empirical papers on ratification of and compliance with the Kyoto Protocol could be used to test our propositions: Fredriksson and Ujhelyi (2006) include the number of 'veto players' as a variable, whereas Jensen and Spoon (2007) explicitly investigate the impact of parties' preferences and their position in government.

References

Alesina, Alberto and Howard Rosenthal (1995) *Partisan politics, divided government and the economy* (Cambridge, Massachusetts: Cambridge University Press)

Anderson, John W (1999) 'A "crash course" in climate change' in Wallace E Oates (ed) *The RFF reader in environmental and resource management* (Washington DC: Resources for the Future), 213–224

Banks, Jeffrey S (1990a) 'Monopoly agenda control and asymmetric information', *Quarterly Journal of Economics*, 105:2, 445–464

Banks, Jeffrey S (1990b) 'A model of electoral competition with incomplete information', *Journal of Economic Theory*, 50:2, 309–325

Banks, Jeffrey S (1993) 'Two-sided uncertainty in the monopoly agenda setter model', *Journal of Public Economics*, 50:3, 429–444

Barrett, Scott (1994) 'Self-enforcing international environmental agreements', *Oxford Economic Papers*, 46, 878–894

Barrett, Scott (1998) 'On the theory and diplomacy of environmental treaty-making', *Environmental and Resource Economics*, 11:3–4, 317–333

Barrett, Scott (2003) *Environment and statecraft: the strategy of environmental treaty-making* (Oxford: Oxford University Press)

Bueno de Mesquita, Bruce, James D Morrow, Randolph Siverson and Alastair Smith (1999) 'An institutional explanation of the democratic peace', *American Political Science Review*, 93:4, 791–807

Cai, Hongbin (2000) 'Bargaining on behalf of a constituency', *Journal of Economic Theory*, 92:2, 234–273

Carraro, Carlo and Dominico Siniscalco (1993) 'Strategies for the international protection of the environment', *Journal of Public Economics*, 52:3, 309–328

Carraro, Carlo and Domenico Siniscalco (1998) 'International environmental agreements: incentives and political economy', *European Economic Review*, 42:3–5, 561–572

Cremer, Helmuth, Philippe de Donder and Firouz Gahvari (2008) 'Political competition within and between parties: an application to environmental policy', *Journal of Public Economics*, 92:3–4, 532–547

Currarini, Sergio and Henry Tulkens (1997) 'Core-theoretic and political stability of international agreements on transfrontier pollution', CORE Discussion Paper 9793, Universitè Catholique de Louvain

Dai, Xinyuan (2002) 'Political regimes and international trade: the democratic difference revisited', *American Political Science Review*, 96:1, 159–165

Dai, Xinyuan (2006) 'The conditional nature of democratic compliance', *Journal of Conflict Resolution*, 50:5, 690–713

Downs, Anthony (1957) *An economic theory of democracy* (New York: Harper and Brothers)

Eckert, Heather (2003) 'Negotiating environmental agreements: regional or federal authority?' *Journal of Environmental Economics and Management*, 46:1, 1–24

Fershtman, Chaim, Kenneth L Judd and Ehud Kalai (1991) 'Observable contracts: strategic delegation and cooperation', *International Economic Review*, 32:3, 551–559

Folmer, Henk, Nick Hanley and Fanny Mißfeldt (1998) 'Game-theoretic modelling of environmental and resource problems: an introduction' in Nick Hanley and Henk Folmer (eds) *Game theory and the environment* (Cheltenham, United Kingdom: Edward Elgar), 1–29

Fredriksson, Per G and Daniel L Millimet (2004a) 'Comparative politics and environmental taxation', *Journal of Environmental Economics and Management*, 48:1, 705–722

Fredriksson, Per G and Daniel L Millimet (2004b) 'Electoral rules and environmental policy', *Economics Letters*, 84:2, 237–244

Fredriksson, Per G and Gergely Ujhelyi (2006) 'Political institutions, interest groups, and the ratification of international environmental agreements', Working Paper, University of Houston, Texas
Gilligan, Michael J (2004) 'Is there a broader-deeper trade-off in international multilateral agreements?', *International Organization*, 58:3, 459–484
Grossman, Gene M and Elhanan Helpman (1995) 'Trade wars and trade talks', *Journal of Political Economy*, 103:4, 675–708
Hammond, Thomas H and Brandon Prins (2006) 'Domestic veto institutions, divided government, and the status quo: a spatial model of two-level games with complete information' in Robert Pahre (ed) *Democratic foreign policy making: problems of divided government and international cooperation* (New York: Palgrave), 21–82
Harrington, Joseph E (1992) 'The role of party reputation in the formation of policy', *Journal of Public Economics*, 49:1, 107–121
Hoel, Michael (1991) 'Global environmental problems: the effects of unilateral actions taken by one country', *Journal of Environmental Economics and Management*, 20:1, 55–70
Hug, Simon (2004) 'Models of multilateral negotiations and ratification', Paper presented at the 2004 annual meeting of the American Political Science Association, Chicago
Hug, Simon and Tobias Schulz (2005) 'Referendums in the EU's constitution building process', *Review of International Organizations*, 2:2, 177–218
Humphreys, Macartan (2007) 'Strategic ratification', *Public Choice*, 132:1, 191–208
Iida, Keisuke (1993) 'When and how do domestic constraints matter? Two-level games with uncertainty', *Journal of Conflict Resolution*, 37:3, 403–426
Iida, Keisuke (1996) 'Involuntary defection in two-level games', *Public Choice*, 89:2, 283–303
Jensen, Christian B and Jae-Jae Spoon (2007) 'Putting the party back in: explaining EU member state compliance with the Kyoto Protocol, 1998–2003', Working Paper, University of Iowa
Kaul, Inge, Isabelle Grunberg and Marc A Stern (1999) *Global public goods* (Oxford: Oxford University Press)
Lizzeri, Alessandro and Nicola Persico (2001) 'The provision of public goods under alternative electoral incentives', *American Economic Review*, 91:1, 225–245
Lohmann, Susanne (1993) 'Electoral cycles and international policy cooperation', *European Economic Review*, 37:7, 1373–1391
Mansfield, Edward D, Helen V Milner and B Peter Rosendorff (2000) 'Free to trade: democracies, autocracies and international trade', *American Political Science Review*, 94:2, 305–321
Mansfield, Edward D, Helen V Milner and B Peter Rosendorff (2002) 'Replication, realism and robustness: analyzing political regimes and international trade', *American Political Science Review*, 96:1, 167–169
Markussen, Peter and Gert Tinggaard Svendsen (2005) 'Industry lobbying and the political economy of GHG trade in the European Union', *Energy Policy*, 33:2, 245–255
Mayer, Frederick W (1992) 'Managing domestic differences in international negotiations: the strategic use of internal side-payments', *International Organization*, 46:4, 793–818
Milesi-Ferretti, Gian-Maria, Roberto Perotti and Massimo Rostagno (2002) 'Electoral systems and the composition of public spending', *Quarterly Journal of Economics*, 117:2, 609–657
Milner, Helen V (1997) *Interests, institutions, and information: domestic politics and international relations* (Princeton, New Jersey: Princeton University Press)
Milner, Helen V and B Peter Rosendorff (1996) 'Trade negotiation, information and domestic politics', *Economics and Politics*, 8:2, 145–189
Milner, Helen V and B Peter Rosendorff (1997) 'Democratic politics and international trade negotiations: elections and divided government as constraints on trade liberalization', *Journal of Conflict Resolution*, 41:1, 117–146
Mo, Jongryn (1994) 'The logic of two-level games with endogenous domestic coalitions', *Journal of Conflict Resolution*, 38:3, 402–422
Mo, Jongryn (1995) 'Domestic institutions and international bargaining: the role of agent veto in two-level games', *American Political Science Review*, 89:4, 914–924
Morrow, James D (1991) 'Electoral and congressional incentives and arms control', *Journal of Conflict Resolution*, 35:2, 245–265

Morton, Sanford (1988) 'Strategic voting in repeated referenda', *Social Choice and Welfare*, 5:1, 45–68
Nordhaus, William and Joseph Boyer (2003) *Warming the world: Economic models of global warming* (Cambridge, Massachusetts: MIT Press)
Pahre, Robert (1997) 'Endogenous domestic institutions in two-level games and parliamentary oversight of the European Union', *Journal of Conflict Resolution*, 41:1, 147–174
Pahre, Robert (2004) 'Formal and informal ratification in the intergovernmental policies of the European Union', Working Paper, European Union Center, University of Illinois, Urbana-Champaign
Pahre, Robert and Paul A Papayoanou (1997) 'Using game theory to link domestic and international politics', *Journal of Conflict Resolution*, 41:1, 4–11
Persson, Torsten (2004) 'Consequence of constitutions', *Journal of the European Economic Association*, 2:2–3, 139–161
Persson, Torsten, Gérard Roland and Guido Tabellini (2000) 'Comparative politics and public finance', *Journal of Political Economy*, 108:6, 1121–1161
Persson, Torsten and Guido Tabellini (1999) 'The size and scope of government: comparative politics with rational politicians', *European Economic Review*, 43:4–6, 699–735
Persson, Torsten and Guido Tabellini (2003) *Economic effects of constitutions* (Cambridge, Massachussetts: MIT Press)
Persson, Torsten and Guido Tabellini (2004a) 'Constitutional rules and economic policy outcomes', *American Economic Review*, 94:1, 25–45
Persson, Torsten and Guido Tabellini (2004b) 'Constitutions and economic policy', *Journal of Economic Perspectives*, 18:1, 75–98
Pollak, Robert A (1998) 'Imagined risks and cost-benefit analysis', *American Economic Review*, 88:2, 376–380
Portney, Paul R (1992) 'Trouble in Happyville', *Journal of Policy Analysis and Management*, 11:1, 131–132
Putnam, Robert D (1988) 'Diplomacy and domestic politics: the logic of two-level games', *International Organization*, 42:3, 427–460
Romer, Thomas and Howard Rosenthal (1979) 'Bureaucrats versus voters: on the political economy of resource allocation by direct democracy', *Quarterly Journal of Economics*, 93, 563–587
Sandler, Todd (1997) *Global challenges: an approach to environmental, political, and economic problems* (Cambridge: Cambridge University Press)
Schelling, Thomas C (1960) *The strategy of conflict* (Cambridge, Massachusetts: Harvard University Press)
Segendorff, Björn (1998) 'Delegation and threat in bargaining', *Games and Economic Behaviour*, 23:2, 266–283
Siqueira, Kevin (2003) 'International externalities, strategic interaction, and domestic politics', *Journal of Environmental Economics and Management*, 45:3, 674–691
Smith, Alastair and David Hayes (1997) 'The shadow of the polls: electoral effects on international agreements', *International Interactions*, 23:1, 79–108
Sunstein, Cass (2006) 'The availability heuristic, intuitive cost-benefit analysis, and climate change', *Climatic Change*, 77:1–2, 195–210
Tarar, Ahmer (2001) 'International bargaining with two-sided domestic constraints', *Journal of Conflict Resolution*, 45:3, 320–340
Tarar, Ahmer (2005) 'Constituencies and preferences in international bargaining', *Journal of Conflict Resolution*, 49:3, 383–407
Tjernstöm, Emila and Thomas Tietenberg (2008) 'Do differences in attitudes explain differences in national climate change policies?', *Ecological Economics*, 65:2, 315–324
Williams, Kenneth C (1994) 'Sequential elections and retrospective voting: some laboratory experiments', *Journal of Theoretical Politics*, 6:2, 239–255

Appendix: Proof of Proposition 3

Note that the curve PO with Pareto-optimal points for the governments is the (only) set of points on which their indifference curves are tangent to each other. Without H's domestic constraints, F is proposing the point where one of its indifference curves is tangent to H's indifference curve that consists of points giving H the same utility as the Nash equilibrium. By comparing the slopes of H's and V's indifference curves at identical points, one sees the new constrained optimum is not on PO any more: $-\frac{B'-C'_H}{B'} < -\frac{B'-\gamma C'_H}{B'}$. Note that V's indifference curves are steeper in every point than H's (for $\gamma > 1$), therefore they have to be tangent with F's concave indifference curves somewhere below PO.

But that does not necessarily mean that at those points at which F's and V's indifference curves are tangent both contribution levels are smaller than at the constrained optimum without domestic constraints. To prove this, we examine comparative statics on the maximization condition for the constrained problem (5), which is:

$$\frac{B'}{B' - C'_F} - \frac{B' - \gamma C'_H}{B'} = 0. \tag{7}$$

If we define the left-hand side of (7) as function G, we can explore if A_H and A_F decrease with an increase in γ:

$$\frac{\partial A_H}{\partial \gamma} = -\frac{\partial G/\partial \gamma}{\partial G/\partial A_H} = -\frac{C'_H}{\frac{B''(B'-C'_F)-B'B''}{(B'-C'_F)^2} - \frac{(B''-\gamma C''_H)B'-(B'-\gamma C'_H)B''}{B'^2}} < 0, \tag{8a}$$

$$\frac{\partial A_F}{\partial \gamma} = -\frac{\partial G/\partial \gamma}{\partial G/\partial A_F} = -\frac{C'_H}{\frac{B''(B'-C'_F)-B'(B''-C''_F)}{(B'-C'_F)^2} - \frac{B''B'-(B'-\gamma C'_H)B''}{B'^2}} < 0. \tag{8b}$$

Even though the contributions of both countries to the international public good are less than those without domestic constraints, H is better off. As we have shown before, H's constraint is not binding when $\gamma > 1$. Therefore, F's new proposal is not on H's indifference curve without domestic constraints—H must have a higher utility compared to the Nash equilibrium.

Environmental security and climate change: analysing the discourse

Maria Julia Trombetta[1]

Introduction: setting the divide

Appeals to 'climate security' represent a recent and fairly successful attempt to introduce environmental concerns into the security agenda. Despite this momentum, the link between environmental change and security remains a contested topic: not only are environmental problems often silenced by more urgent threats but the very opportunity of considering the environment as a security issue is also challenged.

Since environmental problems began to gain relevance in political discourse in the 1970s, there have been a number of suggestions to consider their security implications (Falk 1971; Brown 1977). However, it was only in the 1980s—with the emergence of global environmental problems such as the depletion of stratospheric ozone or global warming—that the debate on environmental security gained momentum. The publication of *Our common future* by the World Commission on Environment and Development (1987) marked the entry of the phrase 'environmental security' into international debates; the merits of the concept were considered on analytical and normative grounds (Soroos 1994).

[1] I would like to thank the four anonymous reviewers for their comments and suggestions.

Environmental security initially appeared to be a good idea, as it was 'meant to alarm traditional security analysts about the issues that "really" matter' (de Wilde 2001, 2) and to increase the relevance of environmental problems in the political agenda. Buzan emphasized that '[e]nvironmental security concerns the maintenance of the local and the planetary biosphere as the essential support system on which all other human enterprises depend' (1991, 19–20). Others welcomed the concept since it 'plays down the values traditionally associated with the nation-state—identity, territoriality, sovereignty—and implies a different set of values associated with environmental change—ecology, globality, and governance' (Dyer 2001, 68). Yet others argued that 'environmental security ... is all about solidarity' (Thompson 1999, 137). On analytical grounds, it seemed a way to provide a better account of new typologies of vulnerability as well as the potential for conflict and violence with which these vulnerabilities could be associated.

Opponents were quick to warn that the term 'security' evokes a set of confrontational practices associated with the state and the military which should be kept apart from the environmental debate (Deudney 1990). Concerns included the possibilities of creating new competencies for the military—militarizing the environment rather than greening security (Käkönen 1994)—or the rise of nationalistic attitudes in order to protect the national environment (Deudney 1999, 466–468). Deudney argued that not only are practices and institutions associated with national security inadequate to deal with environmental problems, but security can also introduce a zero-sum rationality to the environmental debate that can create winners and losers, and undermine the cooperative efforts required by environmental problems. Similar objections came from a southern perspective: environmental security was perceived as a discourse about the security of northern countries, their access to resources and the protection of their patterns of consumption (Shiva 1994; Dalby 1999; Barnett 2001). Although the debate waxed and waned, the concept slowly gained popularity. In April 2007 the security implications of climate change were discussed by the United Nations (UN) Security Council but the state representatives remained divided over the opportunity of considering climate change and, more generally, environmental degradation as a security issue (United Nations Security Council 2007).

The divide between those who oppose the use of the term environmental security by arguing that the logic of security is fixed and inflexible and those who support it by suggesting that the logic of security should be changed[2] distracts attention away from the question of whether practices associated with providing security have been transformed by environmental security discourses. In the literature there is a debate about whether and how security language transforms the method of dealing with an issue—the debate focuses 'on the implications of using security language for the definition and governance of migration and the environment' (Huysmans 2006, 16)—but there is little on the reverse process or on the implications of using environmental language for the definition and

[2] This is the argument of critical security studies. A similar point has been made by Dalby and Barnett who suggested redefining security and environmental security as human security. See Dalby (1998, 2002) and Barnett (2001).

governance of security.[3] This article is an attempt to develop the latter type of analysis by exploring the meaning and function of environmental and climate security. The purpose is to consider how the use of a word in different contexts challenges and transforms the practices and meanings associated with it. It aims to explore 'what the practices of definition and usage *do* to a concept, and what the concept in turn *does* to the world into which it is inscribed' (Bartelson 2000, 182). To undertake this analysis it is necessary to explore how different discourses about environmental and climate security have developed and 'conditioned the possibility of thought and action' (181).

The article is presented in three parts. The first explores why the environment has been excluded from security considerations. By adopting a perspective that is attentive to the social construction of security issues and its implications, the article assesses the potential of a discursive approach in transforming existing security practices. The analysis draws on the theory of securitization elaborated by the Copenhagen School (inter alia Buzan and Waever 1998) and integrates it with elements borrowed from Beck's work (inter alia 1992, 1999, 2006) on risk society to provide a framework that accounts for transformation. It argues that the securitization of environmental issues can reorient security logics and practices. The second and third parts apply this framework to explore the development of environmental security and climate security discourses respectively.

Securitization and the logic of security

The two main arguments against considering the environment as a security issue come from Realists, and from those who warn against the problematic implications the word security brings with it.

Realism—traditionally the mainstream approach in security (or rather strategic studies)—adopts a narrow definition of security that has the state as referent and takes security to be achieved through the menace or use of force. Realists tend to consider environmental problems as belonging to the realm of 'low' politics rather than an issue of 'high' politics, such as security. In this way, as Lacy (2005) has outlined, realists tend to create a hierarchy of threats, distinguishing between threats that can be legitimately included in the security agenda and those that cannot.[4]

Constructivists and poststructuralists have challenged the narrow realist perspective, suggesting instead that threats are socially constructed. The most innovative and thoughtful attempt to conceptualize the social construction of security issues is the theory of securitization as elaborated by the Copenhagen School, a body of research mainly associated with the work of Barry Buzan and Ole Waever. The work of the Copenhagen School is relevant to the ongoing

---

[3] An exception is Dalby (1998). However, Dalby focused on alternative visions of security based on ecological principles rather than exploring the implications of a specific usage in a particular context.

[4] During the Cold War this approach was initially challenged by peace studies and the subsequent emergence of new issues in the economic and environmental field. Nevertheless the intellectual space for a different understanding of security was only opened up by the end of the Cold War. Security, the claim goes, has to be broadened not only to include new threats such as environmental ones, but also to consider other referents below and above the state (Rothschild 1995; Krause and Williams 1996).

analysis because it considers the implications of broadening the security agenda and deals specifically with environmental problems. Moreover it has been influential in the political and academic debate with its warning about the risk of framing environmental problems in security terms.

The theory of securitization argues that there are no objective threats, waiting to be discovered. Instead, various issues can be transformed into security issues if a political community constructs them as such through a successful speech act that transforms the way of dealing with them. Security in this perspective is not a value or a condition but a form of social practice. That is to say, if an issue succeeds in being labelled as a security issue, the method of handling it will be transformed.

The consideration of the discursive formation of security issues provides a new perspective to analyse the environmental security discourse. First, it allows an investigation of the political process behind the selections of threats, exploring why some of them are considered more relevant and urgent than others. Second, it suggests that the awareness of environmental issues can have a relevant role in defining and transforming political communities and their identities, since the process creates new ideas about who deserves to be protected and by whom. Finally, as Behnke (2000, 91) points out, securitization can open the space for a 'genuinely political' constitutive and formative struggle through which political structures (including the practices associated with security) are contested and re-established.

However, for the Copenhagen School, securitization has problematic consequences. The label security brings with it a set of practices and a way of dealing with a problem that characterizes an issue as a security issue. The word security entails a specific logic or rationality, independent of the context or the intentions of the speakers. Security is about survival, urgency and emergency. It allows for exceptional measures, the breaking of otherwise binding rules and governance by decrees rather than by democratic decisions. Moreover security implies a 'decisionist' attitude which emphasizes the importance of reactive, emergency measures. This set of practices is not necessarily codified nor can it be identified by specific rules. Instead it is more a form of rationality, a way of framing and dealing with an issue, or 'a generic structure of meaning which organizes dispositions, social relations, and politics according to a rationality of security' (Huysmans 2006, 24–25). This mindset, once activated, is not open to negotiation. Although it is possible to decide whether or not to securitize an issue—and securitization, as a social process, is determined by a political community rather than by individuals—once an issue is securitized the logic of security necessarily follows.

This logic is borrowed from the Schmittian understanding of the political.[5] For Carl Schmitt (1996 [1932], 37) the political is about the friend-enemy distinction and successfully evoking security brings about that distinction. The logic of security is the logic of war; this suggests an extreme form of antagonism and a zero-sum understanding of security. With the codification and institutionalization of a national security discourse this rationality has been narrowed down to a specific context. Attempts to broaden the security agenda result in the spreading

[5] See Williams (2003) and Huysmans (1999).

of this rationality to other contexts from which it had been excluded (Buzan and Waever 1998).

The Copenhagen School warns about the risk of securitization and distinguishes between securitization—'meaning the issue is presented as an existential threat, requiring emergency measures and justifying actions outside the normal bounds of political procedure'—and politicization—'meaning the issue is part of public policy, requiring government decision and resource allocations' (Buzan et al 1998, 23–24). The School warns that 'when considering securitizing moves such as "environmental security" ... one has to weigh the always problematic side effects of applying a mind-set of security against the possible advantages of focus, attention, and mobilization' (29) and Waever's normative suggestion is: 'less security, more politics!' (Waever 1995, 56).

In the case of the environment this suggestion is problematic. It can lead to the depoliticization and marginalization of urgent and serious issues, while leaving the practices associated with security unchallenged. Many appeals to environmental security have been made not only with the intent of prioritizing issues but also with that of transforming the logic of security and the practices associated with it.

And yet, when applied to environmental issues, the process of securitization does not seem to be analytically accurate. The Copenhagen School, in its empirically driven analysis of various sectors in *Security: a new framework for analysis*, has identified several peculiarities in the environmental sector (Buzan et al 1998, 71–94). Amongst these peculiarities the most noticeable is that few appeals to environmental security have mobilized exceptional measures or inscribed enemies in any context. The Copenhagen School has suggested that when the environment is involved, '"emergency measures" are still designed and developed in the realm of ordinary policy debates' (83). This suggests that issues can be politicized through an appeal to security, a problematic development for the Copenhagen School, which argues that 'transcending a security problem by politicising it cannot happen through thematization in security terms, only away from such terms' (Waever 1995, 56). Even if the School tends to dismiss these as failed securitizations, this seems to show that the transformation of an issue into a security issue can follow different modalities and different logics, which eschew the confrontational logic of the national security model suggested by the School. Through the appeal to security, other logics, which characterize different contexts, can be brought into existence and new actors can gain relevance in security policies.

Securitization, as de Wilde has argued, 'triggers two debates: one about the underlying risk assessment, one about the strategic answer to it' (2008, 596). Successful appeals to security require developing security policies, identifying appropriate strategies and means to deal with the problem. These developments are largely sector dependent and reflect different values, priorities and practices. As Williams has noted, '[s]peech-act theory entails the possibility of argument, of dialogue, and thereby holds out the potential for the transformation of security perceptions both within and between states' (2003, 523).

The emergence of new threats such as environmental problems has suggested that reactive measures and an antagonist understanding of security are not the best ways to deal with these issues. Instead preventive measures appear to be more effective and new means are required. In this context, the growing awareness of the complexity and uncertainty of contemporary threats has suggested to some scholars and policy makers the use of the concept of risk to conceptualize contemporary security dynamics. Themes like risk management and preventive approaches have become more relevant in security discussions.

In this context, Beck's analysis of risk society has been used as 'a means to conceptualise and understand the transformation of Western security policies' (Rasmussen 2001, 285). Beck's analysis of risk society, which has been largely inspired by his analysis of environmental problems and ecological catastrophes, suggests that what he defines as 'risk society' is characterized by a greater number of risks (such as those produced by more complex and dangerous technologies) with new characteristics. Contemporary risks are unbounded and potentially catastrophic. They may affect parts of the globe distant from the place where they originated. For example, nuclear, chemical and genetic technologies have the potential to bring destruction on such a large scale that no remedial action or insurance can be appropriate to compensate for them.

> The security dream of first modernity was based on the scientific utopia of making the unsafe consequences and dangers of decisions ever more controllable; accidents could occur, as long as and because they were considered as compensable. (Beck 2006, 334)

In a risk society instead 'the logic of compensation breaks down and is replaced by the principle of *precaution through prevention*' (334).

According to Beck, a risk society undermines the credibility of institutions and practices dedicated to security provision. First, Beck suggests that it is impossible to disentangle oneself from a web of risk. Contemporary risks affect everybody and it is impossible to create barriers and distance oneself from them—the enemy. For the Copenhagen School security is about the inscription of enemies and the logic of war. On the other hand, Beck points out that '[t]he concept of "enemy" is the strongest possible antithesis to the concept of security' since 'enemy stereotypes empower' as they create 'the relationships and the behavioural logic of attack and defence, pro and contra, which first kill the question and then the people' (1997, 82). Second, Beck challenges the very possibility of having a security logic based on evoking and governing through emergencies. Beck is suggesting that contemporary threats are beyond insurability and '[m]aybe the time has come to work towards the prevention of disorder and catastrophe, and not merely towards their control. Today, there are plans for all kinds of emergencies (ecological, medical, military), but there is no politics to prevent them' (Agamben 2002, 24).

The first consideration targets the antagonistic understanding of security described by the Copenhagen School, whereas the second challenges the neoliberal discourse of risk. This discourse relies on more sophisticated techniques to try to insure even catastrophic risk by shifting it to the capital market. In the case of the environment this discourse is problematic because it can also paradoxically contribute 'to continually generate the condition of emergent catastrophe, in order to profit from it' (Cooper 2004, 8).

Risk society challenges the logic of violence, antagonism and war suggested by securitization. This suggests a set of security practices—based on risk management and on prevention—which are rather different from those suggested by the Copenhagen School. Are appeals to security stuck in fixed problematic practices that the latter suggests? This article claims that through the securitization of nontraditional sectors like the environment, different logics of security can be brought into being. Securitization—broadly understood as the social construction of an issue as a security issue—can be considered as a reflexive process that is not only 'rule-directed' but also 'rule altering' (Beck 1997, 134). Securitization is not about applying a fixed meaning of security as exceptionality that inscribes enemies in a context. Rather, it is 'an always (situated and iterative) process of *generating* meaning' (Stritzel 2007, 366). By securitizing nontraditional issues, the incongruence of a specific logic of security appears while different practices are applied. In this framework, the construction of both threats and rules by which security is carried out are open to a process of social construction and transformation. The following sections explore this process, analysing the development of environmental security and climate security discourses.

Evolving perspectives on environmental and climate security

The emergence of global environmental problems, such as global warming and ozone depletion, resulted in one of the first attempts to securitize the environment on a global scale. The Brandt Report (1980) suggested that 'few threats to peace and survival of the human community are greater than those posed by the prospects of cumulative and irreversible degradation of the biosphere on which human life depends' (quoted in Brauch 2003, 81). These new threats suggested the need to redefine the nature of security in an interdependent world facing new challenges.

In the post-Cold War era, the environmental security discourse opened the window for debating a common approach to security affairs. For instance, at the United Nations General Assembly in December 1988, Gorbachev stressed: 'The relationship between man and the environment has become menacing ... [t]he threat from the sky is no longer missiles but global warming' (quoted in Norman Myers 1993, 11). He also promoted the creation of an Ecological Security Council.[6] As a result of these appeals to security, several initiatives were launched during the early post-Cold War era. Alarming concerns for the hole in the ozone layer transformed the problem into a threat to human health and promoted fairly successful agreements to deal with the issue at an international level. Concerns for climate change not only led to the creation of the Intergovernmental Panel on Climate Change (IPCC)[7] to assess scientific, technical and socio-economic aspects of human impact on climate change but also to the signing of the UN Framework Convention on Climate Change in Rio in 1992 (United Nations 1992). However, negotiations on climate change proved to be more problematic than those on the ozone because acting on climate change would require the transformation of much of the existing economic structure and way of life. Hence climate change was marginalized in the environmental security discourse.

---

[6] The then USSR Foreign Minister Eduard Shevardnadze, at the 43rd UNGA, proposed 'a discussion on how to turn the United Nations Environment Programme into an Environmental Council capable of taking effective decisions to ensure ecological security' (quoted in Nico Schrijver 1989, 118).

[7] The IPCC was established in 1988 by the World Meteorological Organization and by the United Nations Environment Programme.

Despite the initial momentum and the broad scope of environmental security discourses, the debate was captured by discussions about environmentally induced conflicts. This helped both to frame environmental threats in more familiar terms to national security experts and to consider the environment as a legitimate threat. The academic discussion was largely shaped by the work of Thomas Homer-Dixon and a series of well-funded research projects, which aimed to study the relationships between environmental degradation and violent conflicts (Homer-Dixon 1991, 1994). Although Homer-Dixon was cautious in suggesting a straightforward connection between environmental degradation and conflict, his argument was popularized by Kaplan's article 'The coming anarchy' (1994), which drew a grim picture of a future of human misery, migrations and violent conflict, suggesting that the environment would be the 'national-security issue of the early twenty-first century' (58).

The debate about environmentally induced conflicts has since evolved. Research suggests that conflicts are likely to be subnational and low intensity. These results have been corroborated by the projects undertaken by Spillman and Bächler (Bächler et al 1996; Bächler 1999). These projects identified the circumstances conflicts are likely to occur in and emphasized the conflictual dimension of the development process, suggesting that when people face the rapid changes and challenges imposed by the process of development conflicts can follow. Violence is unlikely to occur along the fault-lines between developed and developing countries but it tends to be localized. More pertinently, conflicts were seen as part of the dynamic that draws groups together rather than a disturbance of an otherwise stable society. This suggests that it is not possible to simply condemn conflicts as such (and intervene to restore order), but more emphasis has to be placed on the analysis of when and why different groups decide to resort to violence, and on how to prevent these developments. These results have largely been used by the study 'Environment and Security in an International Context', launched in 1995 by the NATO Committee on the Challenges of Modern Society (CCMS) and carried out by research teams in Germany and in the US (Lietzmann and Vest 1999). The project identified a number of 'syndromes'—sets of complex, pathological relationships between environmental and other social, demographical and political factors—that might help monitoring problematic situations and providing early warning systems for potential conflicts.

Research on environmental conflicts has produced an intense academic debate concerning the empirical validity of the claims, their analytical relevance, methodology and normative implications. The argument that environmental scarcity induces conflicts has been challenged by empirical research showing how environmental degradation often provides an opportunity for cooperation (Hauge and Ellingsen 2001) and, more recently, demonstrating that it is resource abundance rather than scarcity that determines conflicts (Berdal and Malone 2000). Nevertheless this debate is more relevant for the political practices it has determined. During the Clinton administration Homer-Dixon's research was used to promote a more proactive foreign policy (Harris 2001, 121–122). The EU commission, largely influenced by the NATO project, has promoted actions to include environmental considerations into its cooperation for development programmes. The EU has also used environmental concerns to develop and legitimize security capabilities and competencies at the European level.

These capabilities include the creation of a European satellite system for the 'global monitoring for environment and security' (Council of the European Union 2000).

Even if the debate on environmental conflict has been criticized on normative grounds—and some scholars have argued that the debate has represented people in the Third World as 'barbaric Southern Others' (Dalby 1999; Barnett 2001, 67), erased the responsibility of developed countries and tried to frame environmental problems in terms of national security—this debate and the policies it has procured have achieved two things. Firstly, this debate has legitimized new actors and instruments to develop forms of security governance, which downplay the role of the state and of traditional reactive responses. Secondly, it has promoted the development of human security and of a new paradigm of preventive and intervention measures which are often legitimized by the use of the concept.

Duffield and Waddell argued that: '[h]ow conflict has been understood in the post-Cold War period is central to understanding the concept of "security" within human security' (2006, 43).[8] Human security shifted the focus of security from the state to the individual and 'the legitimate concerns of ordinary people who sought security in their daily life' (United Nations Development Programme [UNDP] 1994, 22). The first major formulation of the concept appeared in the UNDP 1994 annual report, which identified environmental security, together with economic, food, health, personal, community and political security as a relevant component of human security and stressed the 'all-encompassing' and 'integrative' qualities of the concept (24). The debate on environmental conflict has played a relevant role in promoting and legitimating forms of intervention, and the development of the concept of 'responsibility to protect' (Duffield and Waddell 2006, 8–10).

The environmental security debate has shown two tendencies. On one hand, the influence of the national security discourse has emphasized the focus on conflicts, transforming the threat to the environment and to the people that depend on it into a threat to global order and stability. Other aspects, such as the impact on health and problems related to over-consumption have been largely ignored in the environmental security discourse. At the same time this discourse has challenged a set of security practices, which focused on military threats and reactive measures and outlined how military responses and preparation are inadequate to deal with environmental issues. And yet it has contributed to shifting the attention to different kinds of vulnerabilities, suggesting that the instruments to provide stability require effort to promote both mitigation and adaptation to environmental impact and change, and that the best results are associated with early intervention and preventive measures.

This is echoed by Duffield and Waddel (2006), who have pointed out the existence of two discourses. The first emphasizes the importance of circulation—it suggests that disasters or conflict in one region have the ability, through population displacement to affect other regions and countries. The second focuses on local consolidation, increasing the resilience of the local population (10). It emphasizes the promotion of diversity and choice and supports the existence of

[8] Even Barnett, who is relatively critical of the environmental conflict thesis, argues that the 'subversive message of Homer-Dixon's work' is that 'environmental problems only have a meaning for security if security is understood in *human* terms' (2001, 64–65).

relationships and institutions able to govern populations. This discourse reflects environmental values and considerations such as diversity and resilience, which are threatened by unsustainable development. This suggests 'a more holistic and ecological view of the security problem, one which sees external threats as a piece of a larger security ecology' (de Lint and Virta 2004, 472). This is not to deny the relationships of power that are deployed in a process of securitization. Instead it suggests that other forms of power—like the discursive and symbolic ones incorporated in the environmental discourse—have played a role in determining a less confrontational approach to security.

These two aspects—one informed by reactive, defensive measures, the other inspired by precautionary ones—have largely been balanced in the environmental security discourse in western countries, even if the balance depended on the different capabilities of the actors involved in the construction of the security discourse. The EU, for instance, has traditionally emphasized a preventive approach over a reactive one. As a result various securitizations of environmental issues have brought about security practices largely inspired by the practices developed within the environmental sector to ensure safety, adaptation and resilience. In this context the 'emergency measures' determined by the appeal to environmental security have been 'designed and developed in the realm of ordinary policy debates' (Buzan et al 1998, 83). Is this development going to be implemented in the recent securitization of climate change, even if it is occurring in the context of the war on terror, which has renewed a confrontational approach to security? To answer this question, it is necessary to consider how the climate security discourse has been framed.

The challenge of climate security

The debate about environmental security has been marginalized in the shadow of the war on terror. The latter has made other threats appear more urgent and serious than the environmental ones. However, over the last few years, concerns over climate changes have gained momentum. Several factors have contributed to this new development: on one hand, there has been a growing consensus on the dimensions of the anthropogenic impact on climate; on the other hand, since the withdrawal of the US from the Kyoto protocol, there have been several 'securitizing moves' aimed at promoting action to counter climate change and its consequences, on both sides of the Atlantic.

In 2004 David King, then UK government's chief scientific adviser, claimed that climate change was a far more serious threat than international terrorism (*BBC News Online* 2004). This was not only a European move. The same year an internal report of the US Department of Defense on the impact of 'an abrupt climate change' was leaked to the press. Drawing a comparison with paleoclimatic events, the report forecasted a rapid climate change: after a decade of slow warming, characterized by severe storms, typhoons and floods—bringing destruction in coastal cities and low-lying islands—the alteration of the Gulf Stream would plunge Northern Europe into a Siberian-like climate and transform key areas in Africa, Asia and Australia into deserts by 2020 (Schwartz and Randall 2003). This report seemed to anticipate the destruction caused by hurricanes Rita and Katrina, which struck the US in autumn 2005. Events that had traditionally been considered as natural disasters were increasingly associated with the growing instability of climate. Moreover they outlined how even the US was not able to protect its own citizens and provide effective disaster relief. Movies like *The day after tomorrow* and *An inconvenient truth* further reinforced the representation of climate change as a threat and a security issue.

These developments followed a typical securitization move by delineating climate change as an urgent, existential threat which demands immediate action. When Margaret Beckett was the UK Foreign Secretary, she promoted an active campaign to prioritize climate change in the international agenda. She was one of the first to employ the term 'climate security' (Beckett 2006). Thus Beckett transformed the debate over climate change into a more dramatic one on climate security, contributing to its securitization, thereby prompting several questions about this securitization and its impact on security practices. Who or what is the object of security in the climate security discourse? Who is supposed to provide security and by what means? What are the implications of the process of securitization of the climate? These questions need to be addressed to analyse the meaning and function of climate security.

Climate security suggests a concern for the security of the climate which is understood as the maintenance of stable climatic conditions as a prerequisite of all human enterprises, rather than the security of the climate itself. Climate security is evoked to secure people and societies that depend on it. As in the case of environmental security, climate security is about 'the maintenance of achieved levels of civilization' (Buzan et al 1998, 76). In this sense, as de Wilde (2008) highlights, climate security captures a paradox because it is the contemporary way of life that is causing environmental problems. Yet to maintain the existing way of life, it is necessary to change many present global structures. The dilemma then becomes one of whether the existing structures should be changed voluntarily or whether it is preferable to wait until 'structural change will be enforced violently and randomly by environmental crises' (595). The problem is even more pressing because action on climate change requires long-term measures: power plants that are built today would last for decades and innovation in the transport system takes time.

This touches the core of the problem and identifies two contrasting approaches to security provisions. The first—the reliance on emergency measures—suggests that it is impossible to be prepared for all potential threats and it is preferable to deal with emergencies as they occur; the second—the development of preventive ones—warns of potential catastrophic impacts. In this sense securitization is about moving from one position to the other. This transformation is evident in the debate on the kind of threats that are posed by climate change and in its recent transformation.

There are two aspects to this transformation. First, much of the debate has been framed in terms of adaptation to climate change, on the assumption that it will be a slow process. This is implicit in the UNFCCC which states that the objective of the Convention is to achieve the

> stabilization of greenhouse gas concentrations in the atmosphere at a level that would prevent dangerous anthropogenic interference with the climate system. Such a level should be achieved within a time frame sufficient to allow ecosystems to adapt naturally to climate change, to ensure that food production is not threatened

> and to enable economic development to proceed in a sustainable manner. (United Nations 1992, article 2)

This discourse assumes the relative stability of climate systems—ongoing changes will occur in a predictable way—and relies on human ingenuity and adaptation capability. This approach has been challenged by the possibility of catastrophic events. That is, when a complex system reaches a tipping point, it can have nonlinear responses, which result in runaway changes that are very difficult to bring under control again. Examples of such catastrophic events would be the disruption of the Gulf Stream, the weakening of the monsoon systems and the instability of continental ice sheets. The recent securitization move is based on this dramatic reframing of the threat.

The second aspect is the shift of importance between two different and overlapping securitizations. The first one identifies climate change as a threat. The second considers environmental policies as a threat. The latter was evident in the US position at the 1992 Earth Summit, when George Bush stated that 'the American lifestyle is not up for negotiation' (quoted in Lerner 1998, 12). The US subsequently withdrew from the Kyoto Protocol, which was considered as a threat to the US economy. This kind of securitization is also evident in the concern expressed by some developing countries, which argued against the imposition of policies aimed at forcing them to reduce their emissions since these policies can threaten their development process. The *Stern review on the economics of climate change* prepared by the economist Nicholas Stern for the British government and published in October 2006 is an intervention in this debate and has contributed to the securitization of climate change instead of that of environmental policy. The *Stern review* focused on the economic consequences of climate change, an aspect that has often been marginalized in the environmental security discourse, and outlined the economic cost of postponing action. In this way the idea that environmental measures can be a threat to economic competitiveness and growth is turned around with a cost-benefit analysis which warned that inaction on climate change could cost the world economy between 5 and 20 per cent of global GDP each year, whereas the cost of reducing greenhouse gas emission to avoid the worst impact of climate change could be limited to around 1 per cent of global GDP annually (Nicholas Stern 2006, vi). The *Stern review* typifies the difference between the European approach—that tends to see environmental policy as an opportunity—and the US one—that sees it as a potential threat.

As demonstrated hitherto, the identification of threats reveals not only the identities, interests and values that are supposed to be protected, but also the practices and logics associated with security. Hence, it is relevant to explore which specific aspects are identified in the climate security discourse, in order to assess its potential and limits in a process of transformation. The list of threats provided by Margaret Beckett is illustrative. She focused on food security, water issues, security of energy supply, immigration, conflicts and failing states, arguing that 'a failing climate means more failing states' (Beckett 2006). In this way, she outlined the complexity of climate change and the variety of people and vulnerabilities involved, and called for a wider and deeper security discourse, which is more attentive to human needs. This approach resonates with the human security discourse and its associated practices. However, what has emerged as the 'threat'—since it is represented as the root cause—in the recent development of the climate security discourse is the emission of greenhouse gases, mainly carbon dioxide. This is evident in the British statement about 2006 strategic priorities, one of which includes 'achieving climate security by promoting a faster transition to a sustainable, low-carbon economy' (Foreign and Commonwealth Office 2007, 70).

A focus on emissions has the merit of involving industrialized countries and avoiding the removal of their responsibility suggested by the discourse on environmental conflicts, which focused on the global south, its inadequacy and responsibility. Moreover, it links the discourse on climate security with that on energy security, suggesting that the two issues are interconnected.

This focus however has two problematic implications. First it tends to reduce the problem of sustainability to one of emissions, creating the illusion that it is enough to simply focus on cutting emissions or switching toward nongreenhouse gas energy, such as nuclear or bio-fuel to solve the environmental crisis. This approach downplays the complexity of environmental problems and the impact of the overuse of limited resources. Instead, climate change is only one aspect of a larger human-made environmental change. As humankind pushes the boundaries of the carrying capacity of the planet, the impact of climate change is going to increase especially on those who live in marginal, vulnerable areas. Reducing emissions will not solve all the other environmental problems or increase the resilience of vulnerable population. Returning to the previous example of hurricane Katrina: even though it has been associated with global warming, its devastating impact was also a result of other local problems, such as the draining of the wetlands or the extraction of groundwater; problems of levee design and maintenance standards; the failure of emergency services which had been focused on other security priorities. There is a trade-off between the focus on reducing emissions and that on improving resilience and adaptability. In this sense, the focus on emissions tends to limit the holistic perspective suggested by environmental security.

The second implication is related to the link between climate security and energy security. This link contains an inherent tension since energy security is traditionally associated with national security and its logic, whereas climate security is supposed to promote a cooperative approach to global issues. Agreements on energy supply are generally the result of bilateral agreements between states. The energy sector contributes a substantial portion of states' income and policy—states gain revenues from concessions, transit fees and taxes or are directly involved with national companies. Hence the current situation of tight energy markets characterized by rising demand, high prices and concerns for terrorist attacks against critical infrastructures is particularly challenging for states, which are divided between promoting privatization and being more involved in energy policy. Besides, there are limited multilateral institutions that deal with energy security, resulting in the higher risks of fostering a zero-sum mentality and an antagonistic attitude which can be problematic in dealing with climate issues.

How can these developments be read through the lens of the framework previously elaborated? Can this be considered as a securitization which is transforming security practices? The renewal of the debate on climate change and security appears as an attempt to transform it into an existential threat, requiring urgent measures. It has mobilized political action, emergency measures and even attempts to institutionalize the debate at an international level. So far the securitization of climate has succeeded in persuading even the reluctant Bush administration to undertake discussion on emissions reduction. It has also contributed to the formulation of the Bali Roadmap to set a strategy for the post-Kyoto period. The UN Security Council, at the initiative of the UK, discussed the potential impact of climate change on peace and security for the first time ever (UK Mission to the UN 2007). The most impressive results have been within the EU, since it has contributed to the EU developing a common energy policy—an issue that has previously been delayed for decades. Traditionally energy issues have been considered a prerogative of member states and security of supply has been considered a national security issue. The EU Commission is promoting a nonantagonistic approach that relies on liberalization and cooperation to promote a common energy policy and to secure energy supply and climate stability. The impact of this strategy is evident in the reaction to the Ukrainian gas crisis in 2006. When Russia cut the gas supply to Ukraine, due to their dispute over gas prices, the amount of gas transiting through Ukraine and destined for European countries fell dramatically (Jon Stern 2006). Despite the rapid solution of the crisis it was considered a wakeup call which prompted a significant debate on energy security. Within NATO the point was discussed in terms of new roles for the alliance, including the possibility of military involvement to patrol the supply routes, suggesting an antagonistic approach (Shea 2006), but within the EU the crisis provided an opportunity to expedite the development of a common energy policy. The common energy policy set ambitious targets, mobilizing consensus through the double lever of climate security and energy security. In January 2007 the Commission presented the 'Energy and Climate package' (Commission of the European Communities 2007). It included a Strategic Energy Review which focused on both internal and external aspects of EU energy policy. In March 2007, EU leaders approved the plan, agreeing on a binding target of 20 per cent reduction of greenhouse gas emissions by the EU by 2020, compared to 1990 levels. Central to the agreement was the recognition that energy and environment should go hand in hand. The plan committed member states to raising the European share of renewable energy to 20 per cent, increasing energy efficiency, completing the internal market for electricity and gas, and the development of a common external energy policy. Although the focus is on the EU interest and security, the means to achieve them are market mechanisms, promotion of liberal order and multilateralism.

Thus far appeals to climate security have mobilized actions even if the emergency measures have not exceeded the ordinary policy debate. Hence these appeals can be considered as proper securitization rather than failed securitizing moves.[9] The securitization of climate change has avoided the identification of enemies and has involved actors other than states, both in the securitizing moves and in the security provisions.

The other relevant aspect of climate security discourse is the securitization of threats that are uncertain, widespread and whose impact is difficult to quantify. The appeal to security is intended to develop precautionary measures, as

[9] The Copenhagen School distinguishes between a securitizing move—a discourse that presents something as an existential threat to a referent object—and securitization—if the audience accepts the discourse (Buzan et al 1998, 25).

Margaret Beckett stated: 'I am in no doubt that today being a credible foreign minister means being serious about climate security because the question for foreign policy is not just about dealing with each crisis as it hits us' (2006). She continued that it is necessary to put 'in place the condition for our future security and prosperity in a crowded and interdependent world'. This is an application of the 'precaution through prevention' approach to security and reflects Beck's argument on risk society (Beck 2006, 334).

The possibility of transforming into a threat something that has not yet materialized and allowing it to bring about the practices suggested by the Copenhagen School in the case of securitization presents a grim perspective. The possible adoption of a precautionary approach to security issues has been criticized on the grounds that it can justify preventive military actions, extensive surveillance measures, the inversion of the burden of proof or actions decided on the worst case scenario (Aradau and van Munster 2007). In the case of the environment, it is possible that the securitization of climate change would result in confrontational politics, with states adopting politics to protect their territory against sea-level rising and immigration; with the Security Council adopting resolutions to impose emission targets, and even military action against polluting factories; and surveillance systems to monitor individual emissions. This possibility, however, depends on taking for granted a security logic based on enemies and extraordinary measures.

What is at stake in the climate security discourse is the possibility of introducing mechanisms to prevent emergencies within a system that tends to rely, on the one hand, on governing through emergencies and, on the other hand, on insurance and compensation. The securitization of climate is an attempt to evoke the symbolic power of an environmental discourse based on interdependence and prevention to establish a framework for security and energy governance at the global level. It is about renegotiating the spaces in which risk management and market mechanisms prevail, and those in which intervention and regulations are legitimated. Securitization remains a very political moment. Its implications largely depend on what is securitized and what means are employed to provide security.

Conclusion

The recognition and constitution of a problem as a threat implies the identification of the political community that deserves protection, the legitimization of the means to provide security and eventually their institutionalization. Some of these aspects are more settled and consolidated than others, as are the different logics of security, such as the antagonistic, emergency-based one evoked by the Copenhagen School. These developed because, within a particular context, they were the most effective response against a specific representation of threats. This, however, does not mean that they are not open to negotiation when challenged by a new environment and threats.

Climate change poses threats that are largely uncertain, diffuse, difficult to quantify and yet potentially catastrophic. This reflects the logic of a risk society portrayed by Beck. This article has explored how the practices associated with security are challenged by the attempts to transform environmental crises and climate change into a security issue, and has shown how appeals to security have emphasized the relevance of preventive, nonconfrontational measures and the importance of other actors than states in providing security. A potentially nonessentialist approach like securitization, which focuses on the social process that specifies threats, can be relevant in studying how various environmental issues gain priority and mobilize social action. However, the Copenhagen School identifies the 'securityness' of security with a specific logic determined by the realist tradition. In this way the School has imposed a problematic fixity that tends to essentialize an historical- and sector-specific understanding of security and the practices legitimized by it. Even if this logic is still relevant, the analysis of environmental security discourses and the securitization of climate change have shown that transforming an issue like climate change into a security issue is not about applying a fixed meaning of security and the practices associated with it. Rather, it is a reflexive and contextualized process that generates meanings and practices.

References

Agamben, Giorgio (2002) 'Security and terror', transl Carolin Emcke, *Theory and Event*, 5:4, 24

Aradau, Claudia and Rens van Munster (2007) 'Governing terrorism through risk: taking precautions, (un)knowing the future', *European Journal of International Relations*, 13:1, 89–115

Bächler, Gunther (1999) *Violence through environmental discrimination* (Dordrecht, The Netherlands: Kluwer)

Bächler, Gunther, Volker Böge, Stefan Klötzli, Stephan Libiszewski and Kurt R Spillmann (1996) *Kriegsursache Umweltzerstörung* [*Environmental degradation as a cause of war*] (Zurich: Rüegger Verlag)

Barnett, Jon (2001) *The meaning of environmental security: ecological politics and policy in the new security era* (New York: Zed Books)

Bartelson, Jens (2000) 'Three concepts of globalization', *International Sociology*, 15:2, 180–196

BBC News Online (2004) 'Global warming "biggest threat"' 9, January, <http://news.bbc.co.uk/1/hi/sci/tech/3381425.stm>, accessed 5 May 2008

Beck, Ulrich (1992) *Risk society: towards a new modernity* (Newbury Park: Sage Publications)

Beck, Ulrich (1997) *The reinvention of politics: rethinking modernity in the global social order* (Cambridge: Polity Press)

Beck, Ulrich (1999) *World risk society* (Cambridge: Polity Press)

Beck, Ulrich (2006) 'Living in the world risk society', *Economy and Society*, 35:3, 329–345

Beckett, Margaret (2006) Untitled speech on foreign policy and climate security given at the British Embassy, Berlin, 24 October, <http://www.britischebotschaft.de/en/news/items/061024.htm>, accessed 5 March 2008

Behnke, Andreas (2000) 'The message or the messenger? Reflections on the role of security experts and the securitization of political issues', *Cooperation and Conflict*, 35:1, 89–105

Berdal, Mats and David Malone (eds) (2000) *Greed and grievance: economic agendas in civil wars* (Boulder, Colorado: Lynne Rienner)

Brauch, Hans (2003) 'Security and environment linkages on the Mediterranean space' in Hans Brauch (ed) *Security and environment in the Mediterranean: conceptualizing security and environmental conflicts* (Berlin: Springer), 35–143

Brown, Lester (1977) *Worldwatch Paper 14: Redefining national security*, (Washington DC: Worldwatch Institute)

Buzan, Barry (1991) *People, states, and fear: an agenda for international security studies in the post-Cold War era* (New York: Harvester Wheatsheaf)

Buzan, Barry and Ole Waever (1998) 'Liberalism and security: the contradiction of the liberal leviathan', Working Paper 23, Copenhagen Peace Research Institute (COPRI), Copenhagen

Buzan, Barry, Ole Wæver and Jaap de Wilde (1998) *Security: a new framework for analysis* (Boulder, Colorado: Lynne Rienner)

Commission of the European Communities (2007) 'An energy policy for Europe', COM(2007)1 final, Brussels, 10 January

Cooper, Melinda (2004) 'On the brink: from mutual deterrence to uncontrollable war', *Contretemps: An Online Journal of Philosophy*, 4, 2–18

Council of the European Union (2000) 'Council resolution on a European space strategy', C371/02, Brussels, 16 November

Dalby, Simon (1998) 'Ecological metaphors of security: world politics in the biosphere', *Alternatives*, 23:3, 291–320

Dalby, Simon (1999) 'Threats from the South? Geopolitics, equity, and environmental security' in Daniel Deudney and Richard Matthew (eds) *Contested grounds: security and conflict in the new environmental politics* (Albany: SUNY Press), 187–219

Dalby, Simon (2002) *Environmental security* (Minneapolis: University of Minnesota Press)

De Lint, Willem and Sirpa Virta (2004) 'Security in ambiguity', *Theoretical Criminology*, 8:4, 465–489

De Wilde, Jaap (2001) 'Environmental security', Unpublished manuscript

De Wilde, Jaap (2008) 'Environmental security deconstructed' in Hans G Brauch (ed) *Globalization and environmental challenges: reconceptualizing security in the 21st century* (Berlin: Springer), 595–602

Deudney, Daniel (1990) 'The case against linking environmental degradation and national-security', *Millennium: Journal of International Studies*, 19:3, 461–476

Deudney, Daniel (1999) 'Environmental security: a critique' in Daniel Deudney and Richard Matthew (eds) *Contested grounds: security and conflict in the new environmental politics* (Albany: SUNY Press), 187–219

Duffield, Mark and Nicholas Waddell (2006) 'Securing humans in a dangerous world', *International Politics*, 43:1, 1–23

Dyer, Hugh (2001) 'Theoretical aspects of environmental security' in Eileen Petzold-Bradley, Alexander Carius and Arpád Vincze (eds) *Responding to environmental conflicts: implications for theory and practice* (Dordrecht, The Netherlands: Kluwer), 67–81

Falk, Richard A (1971) *This endangered planet: prospects and proposals for human survival* (New York: Random House)

Foreign and Commonwealth Office (2007) *Department report: 1 April 2006–31 March 2007* (London: Her Majesty Stationery Office)

Harris, Paul (2001) *International equity and global environmental politics: power and principles in US foreign policy* (Aldershot, United Kingdom: Ashgate)

Hauge, Wenche and Tanja Ellingsen (2001) 'Causal pathways to conflict' in Paul Diehl and Nils Gleditsch (eds) *Environmental conflict* (Boulder, Colorado: Westview Press), 36–57

Homer-Dixon, Thomas (1991) 'On the threshold: environmental changes as causes of conflict', *International Security*, 16:2, 76–116

Homer-Dixon, Thomas (1994) 'Environmental scarcities and violent conflict', *International Security*, 19:1, 5–40

Huysmans, Jef (1999) 'Know your Schmitt: a godfather of truth and the spectre of Nazism', *Review of International Studies*, 25:2, 323–328

Huysmans, Jef (2006) *The politics of insecurity: fear, migration and asylum in the EU* (London: Routledge)

Käkönen, Jyrki (ed) (1994) *Green security or militarized environment* (Aldershot, United Kingdom: Dartmouth)

Kaplan, Robert (1994) 'The coming anarchy', *Atlantic Monthly*, 273:2, 44–76

Krause, Keith and Michael Williams (1996) 'Broadening the agenda of security studies: politics and methods', *International Studies Quarterly*, 40:2, 229–254

Lacy, Mark (2005) *Security and climate change: international relations and the limits of realism* (London: Routledge)

Lerner, Steve (1998) *Eco-pioneers: Practical visionaries solving today's environmental problems* (Cambridge: MIT Press)

Lietzmann, Kurt and Gary Vest (1999) *'Environment and security in an international context'*, *Report 232* (Brussels: NATO Committee on Challenges of Modern Society)

Myers, Norman (1993) *Ultimate security: the environmental basis of political stability* (New York: Norton)

Rasmussen, Mikkel (2001) 'Reflexive security: NATO and international risk society', *Millennium: Journal of International Studies*, 30:2, 285–309

Rothschild, Emma (1995) 'What is security?', *Daedalus*, 124:3, 53–98

Schmitt, Carl (1996 [1932]) *The concept of the political*, transl G Schwab (Chicago: University of Chicago Press)

Schrijver, Nico (1989) 'International organization for environmental security', *Bulletin of Peace Proposals*, 20:2, 115–122

Schwartz, Peter and Doug Randall (2003) 'An abrupt climate change scenario and its implications for United States national security', Global Business Network, Emeryville, <http://www.edf.org/documents/3566_AbruptClimateChange.pdf>, accessed 5 March 2008

Shea, Jamie (2006) 'Energy security: NATO's potential role', NATO Review, 2006:3, <http://www.nato.int/docu/review/2006/issue3/english/special1.html>, accessed 5 March 2008

Shiva, Vandana (1994) 'Conflicts of global ecology: environmental activism in a period of global reach', *Alternatives: Global, Local, Political*, 19:2, 195–207

Soroos, Marvin (1994) 'Global change, environmental security, and the prisoner's dilemma', *Journal of Peace Research*, 31:3, 317–332

Stern, Jon (2006) 'The Russian-Ukrainian gas crisis of January 2006', Energy Comment (Oxford: Oxford Institute for Energy Studies)

Stern, Nicholas (2006) *The Stern review on the economics of climate change* (London: HM Treasury)

Stritzel, Holger (2007) 'Towards a theory of securitization: Copenhagen and beyond', *European Journal of International Relations*, 13:3, 357–383

Thompson, Michael (1999) 'Security and solidarity: an anti-reductionist analysis of environmental policy' in Frank Fischer and Maarten A Hajer (eds) *Living with nature: environmental politics as cultural discourse* (Oxford: Oxford University Press), 135–150

UK Mission to the UN (2007) 'Energy, security and climate', UK Concept Paper, 23 April, <http://www.britishembassy.gov.uk/servlet/Front?pagename = OpenMarket/Xcelerate/ShowPage&c = Page&cid = 1183543073739>, accessed 5 March 2008

United Nations (1992) 'Framework convention on climate change', FCCC/INFORMAL/84 GE.05-62220 (E) 200705, 9 May

United Nations Development Programme (1994) *Human development report 1994* (New York: Oxford University Press)

United Nations Security Council (2007) 'Security Council holds first-ever debate on impact of climate change', SC/9000, 17 April

Waever, Ole (1995) 'Securitization and desecuritization' in Ronnie Lipschutz (ed) *On security* (New York: Columbia University Press), 46–86

Williams, Michael (2003) 'Words, images, enemies: securitization and international politics', *International Studies Quarterly*, 47:4, 511–531

World Commission on Environment and Development (1987) *Our common future* (Oxford: Oxford University Press)

Securitizing climate change: international legal implications and obstacles

Shirley V Scott[1]

Introduction

In April 2007 the United Nations Security Council (UNSC) held an open debate on climate change, energy and security. While this might at first glance seem to have confirmed the successful securitization of climate change at the international level, this is far from the case. There is a politics of securitization. Both critics and advocates of referring to climate change as a security—as opposed to solely an environmental—issue acknowledge that one objective of doing so is to raise the profile of the issue and the resources and expertise that governments allocate to it. At an international level there are, however, other implications, including that of the relevant institution and body of international law that can be invoked in addressing the issue. This article identifies three conceptualizations of climate change as a security threat. It does not aim to evaluate the appropriateness of those conceptualizations, nor to construct a normative argument in favour of one or the other. Rather, it outlines the international law ramifications of each and how those ramifications might impact on the international politics of climate change were that conceptualization to become widely accepted. It draws on this analysis to illuminate why it is that climate change has not yet been, and is unlikely to be, fully securitized in world politics.

---

[1] The author would like to acknowledge the valuable research assistance of Tim Aistrope and Alexandra Walker.

The politics of securitization

The term 'security' is in contemporary discourse frequently prefaced by a wide range of descriptors, including food, water, retail, web and oil. The process through which members of a community come to accept the juxtaposition of a particular phenomenon or issue with 'security' can be referred to as securitization. One of the most fundamental points emphasized by scholars of securitization has been that it is not the inherent qualities of an object that determine whether it comes to be regarded as a security issue. Securitization is a self-referential practice; it is through presenting an issue as an existential threat and the audience accepting the existence of such a threat that the issue becomes a security issue (Buzan et al 1998, 24). This was exemplified through a debate that took place within the international relations discipline centred on North America and the United Kingdom (UK) in the late 1980s and early 1990s regarding the merits of referring to environmental challenges as security issues (see, for example, Deudney 1990; Levey 1995; Matthews 1989). This was not, in the main, a debate regarding whether environmental degradation was a serious matter to be addressed by policy-makers but a debate as to whether it was appropriate and useful to label it a security issue.

Richard Falk had in 1971 been the first person to argue systematically that environmental change is a security issue (Barnett 2003, 8). A landmark so far as official acceptance of the term at an international level is concerned, was its incorporation into the report of the World Commission on Environment and Development, *Our common future*, in 1987. References to environmental and ecological security became much more widespread in the context of discussion of a 'new security' agenda for the post-Cold War era. Advocates of new ways of thinking about security proposed that, just as the concept of national security had been expanded in the 1970s to include international economics, so should it be broadened further to include resource, environmental and demographic issues (Mathews 1989). The term 'environmental security' was, however, eclipsed by discussion of other matters at the Earth Summit in 1991 (Dalby 1997, 15). Although the term 'environmental security' is no longer radical, it would be fair to say that it remains contested and has not transformed the practice of international institutions and diplomacy to the extent that its advocates might have hoped. The term 'environmental security' has thus never been fully securitized, in either the academic community or in world politics.

Debate regarding the appropriateness of referring to environmental change as a security threat provides a useful context within which to view the current debate regarding the merits of conceptualizing climate change as a security threat. One of the earliest efforts to link the specific environmental issue of climate change to security was Lester Brown's *Redefining national security*, published in 1977 (Barnett 2001, 3). After that—and despite climate change developing into one of the most prominent environmental issues—the literature on environmental security for many years contained little systematic analysis as to just how climate change constituted a threat to security (Barnett 2001). Momentum towards acceptance of climate change as a national security threat has grown in many countries over the last two to three years. In the United States (US), the ongoing socio-economic as well as environmental legacies of Hurricane Katrina gave Americans a visual image of what climate change might mean for the future, even though this particular emergency could not be directly attributed to climate change. Public debate in the US shifted from the economics of climate change to the security implications of climate change (Busby 2007, 1).

In several countries, public officials with responsibilities for security and policing have made comments to the effect that climate change will pose one of the most important challenges in the years to come. The Australian Federal Police Commissioner, Mick Keelty, declared in September 2007, for example, that climate change could turn border security into Australia's biggest policing issue as millions of displaced people look for new land (*ABC News* 2007). A number of nongovernmental, governmental and intergovernmental bodies released reports during 2007 examining the security implications of climate change and the International Institute for Strategic Studies identified climate change as a threat in its *2007 Strategic survey* (Nicholl 2007). In October 2007 the Nobel committee awarded former US vice-president Al Gore and the Intergovernmental Panel on Climate Change its peace prize.

The UK has engaged in a diplomatic push to have climate change securitized at an international level. One of its objectives has undoubtedly been to increase the profile of the perceived climate crisis and the seriousness with which it is treated. John Ashton, who has since June 2006 been the UK Foreign Secretary's Special Representative for Climate Change, has openly admitted that shifting climate change to the security agenda was done to increase the political clout of those arguing for urgent action: 'while environment experts know how to make emissions trading work, it's a "political fact" that you get a quicker response to a security crisis' (Ashton 2006). Securitizing climate change would bring assets to the disposal of those addressing climate change, including the 'know-how to build international coalitions, and the kind of influence over governmental decision-making that environment ministers can only dream of' (Howden 2007). Margaret Beckett, the British Foreign Secretary between May 2006 and June 2007, gave a speech at the British Embassy in Berlin on 24 October 2006, in which she argued strongly that climate change is a threat to international security (Beckett 2006).

The UK went on to use its April 2007 presidency of the UNSC to propose that the Council hold its first debate on climate change and conflict. Fifty-five countries spoke at the daylong meeting, though the Council did not adopt any statement or resolution. The debate revealed deep opposition to the securitization of climate change on the part of China and many developing countries, and implicit rejection of the concept by the US, making it unlikely that, outside the academy, climate change will at the international level be fully accepted as constituting a security threat. Given that developing countries stand to suffer more than most developed countries from the effects of climate change, the opposition of developing countries to the securitization of climate change—as revealed during the UNSC debate—would seem to suggest that it is unlikely to be the profile-enhancing, increased-funding, political implications of securitizing climate change to which they are opposed.

A second, related reason that has sometimes been offered by critics of securitizing environmental issues at the national level is that this will engage the wrong institutions. National security organizations may not have the appropriate resources or expertise because they are designed for other purposes. Critics claim that environmental security tended to direct focus to environmental causes of conflict rather than broader questions of human security (see Brock 1997;

Barnett 2001, 11). '[D]efining environmental issues in terms of security risks is in itself a risky operation. ... we may end up contributing more to the militarization of environmental politics than to the demilitarization of security politics' (Brock 1992, 98). According to Deudney (1990, 467), the institutions tasked with ensuring national security focus on near-term objectives and are deeply committed to zero-sum thinking; hence, they are not the appropriate agencies to deal with environmental problems. At the international level the principal mechanism by which the international community has addressed the issue of climate change has been through the United Nations Framework Convention on Climate Change (UNFCCC)[2] and its Kyoto Protocol.[3] The institution with primary responsibility for international peace and security, on the other hand, is the UNSC. The most obvious institutional ramification of securitizing climate change would appear to be that of opening up a space within which the UNSC could play a role in the international response to climate change.

Climate change as a security threat and the Security Council

The UNSC may make suggestions regarding the peaceful resolution of international disputes and has over the last decade placed considerable emphasis on developing its conflict prevention role, but the most significant aspect of the Security Council entering into the international politics of climate change would be if it were to draw on its Chapter VII powers to make decisions with which states were legally bound to comply.[4] This contrasts with the multilateral treaty-making process typically used to address environmental issues, participation in which is optional. If the UNSC were to make a decision in relation to climate change, the resulting obligations would supersede other legal obligations for the states concerned.[5] In order for the Security Council to take a decision legally binding on states, it must first of all identify a 'threat to the peace, breach of the peace, or act of aggression'.[6] Academic literature and reports by think-tanks and UN bodies have identified two different senses in which climate change constitutes a threat to international peace and security. Let us review both before seeing whether they would be adequate to trigger a Chapter VII response by the UNSC.

The first way that climate change is sometimes portrayed as a security threat is as a catalyst for violence and war. A case is often made that climate change can or could cause violent conflict. Climate change may, for example, lead to desertification, and hence fighting may break out over access to inadequate supplies of water. The violence may be anything from local skirmishes, to an internal breakdown

[2] United Nations Framework Convention on Climate Change, 1992, reprinted in 31 ILM 849 (1992).

[3] The Kyoto Protocol to the United Nations Framework Convention on Climate Change, 1998, reprinted in 37 ILM 22 (1998).

[4] Article 25 of the UN Charter: 'The Members of the United Nations agree to accept and carry out the decisions of the Security Council in accordance with the present Charter.'

[5] Article 103 of the UN Charter states that, 'In the event of a conflict between the obligations of the Members of the United Nations under the present Charter and their obligations under any other international agreement, their obligations under the present Charter shall prevail.'

[6] Article 39 of the UN Charter.

of law and order, to terrorism or interstate conflict. Climate change may either be the primary cause of the conflict, or may exacerbate existing socio-economic tensions. Other factors such as ethnic conflict or poor governance may be as significant or more significant (see Podesta and Ogden 2007–2008, 129). A conceptualization of climate change as a cause of violent conflict is most frequently drawn upon in making a case for climate change being accepted as a security threat. This is probably because, presenting climate change as a cause of violent conflict and war is not departing significantly from a traditional understanding of threats to security but identifying climate change as one cause of conflict.

A study by International Alert, released in 2007, produced a list of 46 states facing a high risk of armed conflict as a knock-on consequence of climate change. These included Afghanistan, Bosnia and Herzegovina, Guinea Bissau, Iran, Iraq, Lebanon, Peru, the Solomon Islands and Sri Lanka. The list is notable because many of the 46 states are currently or have recently been affected by violent conflict. A second list showing 56 states facing a high risk of political instability as a knock-on consequence of climate change includes Albania, Azerbaijan, Brazil, Egypt, Fiji, Laos, Montenegro, Serbia (Kosovo), Thailand, Turkey and Yemen. The report emphasizes that, for these states, insecurity cannot be avoided by reducing carbon emissions; what is essential now is for planning to adapt to the challenges of climate change (Smith and Vivekananda 2007, 4, 44). An influential group of retired US generals and admirals released a study in 2007 entitled *National security and the threat of climate change*, in which they referred, amongst other things, to the potential for extremists to exploit the unstable conditions brought about, or exacerbated by, climate change (CNA Corporation 2007). Air Chief Marshall Sir Jock Stirrup, Britain's military chief, said in a speech in June 2007 that the political, economic and social deprivation induced by climate change could trigger attacks akin to those of 11 September 2001. Indeed, US noncompliance with the Kyoto Protocol had been cited by Osama Bin Laden as a justification for terrorism (Chatham House 2007).

Although a number of analysts question whether climate change is likely to be the principal cause of violent conflict (see Buckland 2007, 1; Nordas and Gleditsch 2007), few would deny that the most severe impacts of climate change might contribute to the outbreak of future conflicts. The United Nations Environment Programme released a report in June 2007 showing that climate change has been one factor in the complicated conflict in Darfur. A number of other factors—including political, religious, ethnic and economic, as well as tribal and clan divisions, land tenure deficiencies and historical feuds—have contributed to tensions in Sudan. Competition for pastoral land and water has long been a driving force behind a number of local confrontations in Darfur, but has become more difficult to contain (United Nations Environment Programme 2007, 85).

A second way of conceptualizing climate change as a security threat regards climate change as a security threat because it will impact directly on human life and well-being. At the national level it has been argued, for example, that extreme weather events could severely damage critical military bases in the US or industrial production in China, with implications for the global economy (Busby 2007). Tuvalu may need to be abandoned if it goes under the sea, and large, heavily populated areas of Bangladesh might be flooded. Elsewhere, melting glaciers may form unstable lakes prone to flooding, which may drown local communities and then leave the area without adequate supplies of fresh water (Paskal 2007, 7). In this conceptualization of climate change as a security threat, the threat to human life and well-being is coming direct from nature—from higher temperatures, extreme weather events or reduced availability of drinking water for example—even if the environmental change was caused by human activity. This differs from the first conceptualization above in that the threat is not from violent people, or even from people in need such as refugees, but from our environment. If such scenarios were considered isolated natural incidents, the rhetoric might be that of 'natural disasters'. But experts are painting a picture of widespread devastation and chaos if climate change cannot be dramatically slowed. It may not only be poor people in developing countries who are affected; there may be crises in such major western cities as London and New York (Paskal 2007).

Either of the conceptualizations of climate change as a security threat would seem to be sufficient to warrant the UNSC identifying a 'threat to the peace', the article 39 prerequisite to a Chapter VII decision on climate change on the part of the UNSC. The war-time United Nations planners may well have had in mind the invasion of one country by another as the principal threat to international peace and security, but article 39 does not use that language. Some lawyers believe that there are limits to the discretion of the UNSC to determine that a threat to peace exists (see, for example, De Wet 2004, 176), but many would agree that in practice a threat to the peace is whatever the Council deems it to be. Since the end of the Cold War, the Council has broadened its understanding of the term, notably in reference to internal crises, to terrorism, and to the proliferation of weapons of mass destruction and their means of delivery. Although a conceptualization of climate change as a security threat because it causes wars is that closest to a traditional understanding of the role of the Council, a Council could also respond to climate change as a direct security threat if it determined the existence of a threat to the peace.

The UNSC would not, then, need to wait until climate change had prompted war in order to act. The UNSC can authorize a response to an actual or threatened internal conflict or humanitarian crisis and could do so if the threat to a specific country were of a direct nature. Indeed, it is possible that the UNSC has already responded to situations made worse by climate change. Drought was a factor in the crisis in Somalia to which the UNSC on 3 December 1992 responded with a resolution authorizing 'Operation Restore Hope' (Murphy 1998–1999, 1184). Much more controversial is the question as to whether humanitarian intervention without UNSC permission is ever legal. Although it is possible to make an ethical case for intervention in humanitarian emergencies if no other alternative is available, the primary argument against legalizing humanitarian intervention without UNSC authorization is the risk of abuse and its offering legal cover for an intervention motivated by considerations of *realpolitik* (Matheson 2006, 139). Debates regarding the legality of humanitarian intervention are not fundamentally affected by acceptance of the view that climate change may cause or exacerbate domestic tensions, or that it may directly cause humanitarian crises, because it is the humanitarian crisis itself to which an intervention responds, not the underlying causes of that crisis.

What actions could the council take in response to the threat of climate change?

If the UNSC identifies a threat to the peace, breach of the peace or act of aggression, the Council has the authority under article 39 to make recommendations, or to decide what measures shall be taken in accordance with articles 41 and 42, to maintain or restore international peace and security. By article 40, the Council may issue provisional measures to prevent an aggravation of the situation. By article 41, the Council may decide what measures not involving the use of armed force are to be employed to give effect to its decisions and it may call upon the members of the United Nations to apply such measures. Should the UNSC consider that the measures provided for in article 41 would be inadequate or have proved to be inadequate, it may go on under article 42 to 'take such action by air, sea, or land forces as may be necessary to maintain or restore international peace and security'. Although the original intention was for states to make agreements with the United Nations regarding the supply of military personnel for such actions, the understanding of article 42 has evolved to mean that the UNSC authorizes member states to take the necessary military action.

The notion of climate change as a cause of interstate conflict sits comfortably with the international law framework centred on the UN Charter. The current international law regime concerning when force may legally be used is based on the notion that using force to achieve political objectives is wrong. The general principle embodied in article 2(4) of the United Nations Charter is that no state is allowed to threaten or to use force against another state. If there is an act of aggression, the attacked state is permitted to exercise its 'inherent' right of self-defence and it may request assistance from other states. But it is to report its act of self-defence to the UNSC, the assumption being that the Council will then take over. The system of collective security established by the United Nations Charter does not delve into the motivations behind a state threatening to use force or using force against another state. Few in the Western world could, for example, explain the history behind Saddam Hussein's invasion of Kuwait in 1990 and why the Iraqis might have considered the invasion legitimate. There are generally multiple causes of war or conflict, and it is quite possible that environmental factors have often been among those causes (Homer-Dixon 2001). The UNSC could, having identified climate change as a threat to international peace and security, authorize the use of force in response. One ground for opposing the securitization of climate change has been that a military response is not appropriate for climate change and its effects (Buckland 2007), but a threat to international peace and security does not have to be met by a military response. The Council could, for example, require that some or all states take action to assist a state such as Bangladesh mitigate the impacts of climate change and adapt to altered geographical realities.

The distinction between direct and indirect security threats is not as significant to the type of action that the Council could require of states as that which has developed through recent practice of the Council, between specific threats at a particular time and in a particular place and general threats of concern to everybody. Conceptualizing climate change as a general cause of violence and conflict or as a widespread, direct threat to international security has some similarities with the Council's recent approach to other threats such as terrorism or weapons of mass destruction (WMD). Beginning with resolution 1373 of 28 September 2001 relating to international terrorism, and continuing with resolutions 1422 and 1487 on the International Criminal Court, and 1540 on WMD, the UNSC has used its Chapter VII powers to require all states to take or not take specific actions in a context not limited to disciplining any particular country (Szasz 2002, 901). Resolution 1373 (2001) contained a 'detailed and seemingly exhaustive list of obligations that states should undertake in the collective struggle against terrorism' (Luck 2006, 103) and established the Counter-Terrorism Committee to monitor the resolution's implementation. In resolution 1540 (2004) the Council drew on its Chapter VII powers to decide, *inter alia*, that states were not to support efforts of nonstate actors to access nuclear, chemical or biological weapons and their means of delivery; that states were to legislate at a national level to prohibit nonstate access to these weapons; and that states were to adopt and enforce effective domestic controls to, *inter alia*, develop and maintain appropriate border controls and law enforcement efforts to prevent illicit trafficking in such items. Resolution 1540 also established a committee to receive reports from states and states were 'called' to report on measures taken to implement the resolution.

The recent legislative activity of the UNSC has drawn criticism from some countries, including Germany, on the grounds that such activity should be left to more representative bodies such as the General Assembly or Economic and Social Council (ECOSOC). This has in turn prompted debate on the part of international lawyers as to the limits of what the UNSC can require of states under its Chapter VII powers and whether decisions of the Council are subject to judicial review by the International Court of Justice. The most basic limit on the power of the UNSC is the procedural limitations deriving from the Charter rules on voting, as contained in article 27. In practice, these mean that a substantive decision requires that no permanent member veto the decision and that the decision enjoys the support of at least four of the ten nonpermanent members. Beyond this, it appears that the scope of the powers of the Council were deliberately left quite open-ended so as to permit it to take whatever action might be required to address any crisis that might arise (Luck 2006, 23). Indeed, the question of the legal limits on the Chapter VII powers of the UNSC lies precisely at the intersection of law and power politics, just at the point where politics supersedes law. If the International Court were to determine, for example, that the Council had made an illegal decision and the Council chose to ignore the Court, the reputation of the system of international law would suffer, with no clear winner.

Two other political limitations can be readily discerned, relating to the perceived political legitimacy of Council decisions. The first is that the action decided upon must be necessary. The 'peoples' at San Francisco were prepared to permit the permanent five (P5) members of the UNSC to assume such far-reaching powers only so long as the action decided on by the Council is genuinely necessary. A second key aspect of the legitimacy of Chapter VII decisions is that they be perceived as in the interests of all. In conferring primary responsibility for the maintenance of international peace and security on the Council, UN members agreed in article 24(1) that in carrying out its duties the UNSC acts on their behalf. Many might consider that the US attempt to have the UNSC authorize the 2003 invasion of Iraq was inappropriate insofar as that the US determination to oust Saddam Hussein was seemingly motivated by the US national interest rather than to ensure international peace and security for all.

This raises the possibility of the Council overseeing the implementation of mandatory measures to address the causes and consequences of climate change. The UNSC becoming the peak body addressing climate change would have seemed unthinkable even a few years ago, but there is increasing recognition that it is too late to prevent climate change. Indeed urgent action is required to prevent the most extreme and dangerous climate change and to mitigate and adapt to climate change and associated developments. It is too late to undertake experiments or make mistakes. There are a number of well-documented problems of the multilateral treaty approach to addressing serious issues of concern to all that make it feasible to at least consider alternatives. Problems common to most multilateral environmental agreements and evidenced in the Kyoto Protocol include the fact that participation is optional and the lowest common denominator problem—in reaching agreement on the laws to which they are going to subject themselves, states are forced to settle for that acceptable to the least willing (Kellow forthcoming). The Kyoto emissions reduction target is inadequate even if achieved (Intergovernmental Panel on Climate Change 2007, 32). There is also the challenge of ensuring compliance. The bottom line is that, while multilateral treaties do impact on state behaviour, they rarely solve the problems they were negotiated to address (Scott 2004).

The Bali Conference demonstrated recognition of what is needed from the next major global treaty on climate change and took other important steps such as establishing an Adaptation Fund, but this does not guarantee that the negotiations on the details will not be dogged by the 'I'm not giving more than you' syndrome. There is, for example, no mandate for the negotiations to set binding targets to ensure adequate funding for the Adaptation Fund (Oxfam 2007), and it is quite possible that no substantial progress will be made on the new treaty until there is a new Administration in the White House. Given the urgency of addressing climate change and the recognized deficiencies of the multilateral treaty approach, a less democratic, more effective method of global regulation may soon become a necessary alternative, just as a more interventionist, centrally planned approach by governments has been mooted in national debates (see, for example, Dunlop 2007; Flannery 2007; Gow 2008; Monbiot 2006). Perhaps the Council could establish an Environment Security Committee to monitor compliance with binding international emissions commitments and require states to report their actions to that committee (see Penny 2007, 61). Although the Council would have the legal authority to override the existing UNFCCC-Kyoto framework, it could just as easily work in conjunction with that framework and lend its weight to making that system more effective.

Given the dire predictions of the consequences of climate change, it is quite reasonable to think that global legislation may in the medium term come to be perceived as necessary—but it is at this stage difficult to see a time at which there would not be procedural obstacles to doing so or when developing countries might regard legislation by the Council as in the interests of all. Article 3(1) of the UNFCCC notes that developed country parties should take the lead in combating climate change and the adverse effects thereof. Both the UNFCCC and the Kyoto Protocol incorporate the principle of common but differentiated responsibilities (CDR) and developing countries have interpreted rhetoric on climate change and security as a move away from that principle. Public opinion polls show that large majorities around the world support taking action to address the problem of global warming and believe that action should be taken by both developing and developed countries, but also that the developed countries should provide developing countries with aid to do so (WorldPublicOpinion.Org 2007). The Bali Action Plan does not explicitly exclude the possibility of developing countries assuming commitments, as did the 1995 Berlin Mandate, but developing countries want the developed world to show a genuine commitment to change on the scale needed to make a difference. The Council would only be able to find a way forward if there were political accommodation between China and the US. The US is not opposed to legislative behaviour on the part of the UNSC *per se*, so long as this can be achieved without any ceding of its own sovereignty, but the US has lent no support to the securitization of climate change. China has been a leading voice in claims that the Council is not the appropriate body to address this issue. Neither has been willing to lead on climate change.

A conceptualization of climate change as a general security threat does, then, have the potential to bring about significant changes in the regulatory approaches adopted to address the climate change issue at a global level. It does nevertheless need to be borne in mind that, even if there are problems with the effectiveness of the multilateral treaty approach, this does not guarantee that leadership by the UNSC would be successful. The initial reports received by the Counter-Terrorism Committee were encouraging, but there has been a drop-off in response rates. Not all states have the capacity to prepare the necessary reports but, even more worrying are those states that appear to have the capacity but not the will to report (Luck 2006, 106). This raises the question as to how the Council would enforce resolutions establishing global standards, expectations, reporting and assessment on climate change matters. In the counterterrorism and nonproliferation areas, the Council has invoked Chapter VII with great frequency but 'blinked' in the face of the challenge of North Korea (Luck 2006, 109). Even if the UNSC were to require climate change actions by all states and to provide capacity-building for those unable to comply, what could or would it do in the face of blatant defiance?

The international legal implications of securitizing climate change as comparable in effect to WMD

One variant of the conceptualization of climate change as a direct security threat is that which compares its effects to those of WMD. John Houghton, who was Head of the UK Met Office before becoming co-chair of the UN intergovernmental panel on climate change, commented publicly in 2003, that the long-term security of the UK is threatened by a problem at least as dangerous as chemical, nuclear or biological weapons: human-induced climate change (Houghton 2003). Speaking to the UK Royal Society in 2005, its President, Lord May, drew attention to the increasing incidence of extreme weather events, the serious consequences of which are rising to levels which invite comparison with WMD (Royal Society 2005). In 2006 then UN Secretary-General Kofi Annan demanded that world leaders give climate change the same priority as they did to curbing the spread of WMD (Egelko 2006). At an African Union summit in early 2007 Yowri Museveni, President of Uganda, referred to climate change as an act of aggression by the rich world against the poor and demanded compensation (The Economist 2007). Mike Townsley of Greenpeace has simply referred to 'weather of mass destruction' (Clark 2007).

As with other conceptualizations of climate change as either a direct or indirect security threat, this conceptualization could provide a basis for the Council to make a Chapter VII decision. But this variant of climate change as a direct security threat has another legal significance insofar as weapons usually inflict damage through the act of attack. 'Attack' is a term used in international law in the context of 'armed attack' giving rise to a legal right of self-defence. A situation necessitating individual or collective self-defence is the only one in which a state may unequivocally use force other than when authorized by the UNSC to do so. To express it slightly differently, this is the only legal justification other than that of UNSC authorization a state can marshal to account for using force. Principle 21 of the Stockholm Declaration, subsequently reaffirmed in principle 2 of the Rio Declaration and widely regarded as constituting customary international law, provides that States have a responsibility to ensure that activities within their jurisdiction or control do not cause damage to the environment of other States or of areas beyond the limits of national jurisdiction. A state may defend itself in the event of an armed attack and it may request assistance from other states to do so. International law already incorporates an allocation of primary responsibility for climate change. The preamble to the UNFCCC notes that 'the largest share of historical and current global emissions of greenhouse gases [GHGs] has originated in developed countries'; this is reinforced in article 3(1), which acknowledges that because of differentiated responsibilities, developed country Parties should take the lead in combating climate change and the adverse effects thereof.

The conceptualization of climate change as comparable in effect to WMD combined with the current legal regimes on climate change and use of force could provide a rhetorical basis for one or more developing countries to justify use of force against developed country emitters of GHGs. It is, however, highly unlikely that an interpretation of 'armed attack' to encompass attack via emissions of GHGs would gain the widespread acceptance necessary to bring about change in customary international law. After September 11, the UNSC made clear its acceptance of the September 11 terrorist attacks as armed attacks. This was generally acquiesced in by the international community, but it is at this stage difficult to envisage comparable acceptance of GHGs emissions constituting armed attack. Secondly, it is difficult to imagine the circumstances in which it would benefit a victim of such an attack using force or even to know who in particular to attack in response. 'No weapon system can halt the advance of a hurricane bearing down on a city, or stem the rising sea, or stop the glaciers melting' (Ashton 2006). Related to this, and perhaps most fundamental of all, is the fact that states such as Bangladesh, that are expected to suffer the worst consequences of a climate change 'attack', may not be in a position to use force against those who have most historical responsibility for climate change.

The international politics of climate change is evolving rapidly, and if far-reaching regulations relating to emissions reductions, mitigation strategies and adaptation were adopted, the question could arise as to whether self-defence might be exercised against a recalcitrant state not complying with the regulations or whether self-defence against a state not complying with the relevant laws might one day be given as justification for an attack motivated by other factors. Although the chain of reasoning still contains some weak or missing links, a conceptualization of environmental changes emanating from global warming as an attack akin to that by WMD has potentially significant legal implications, although the practical political significance is probably not great, at least in the immediate future.

The UN Security Council open debate on energy, security and climate, 17 April 2007

Having reviewed the international legal and institutional implications of securitizing climate change, we are now better placed to understand the positions adopted during the UNSC debate of 17 April 2007. In preparation, the UK on 5 April issued a concept paper setting out the aims of the debate, issues to be considered and possible questions for discussion (UK Mission to the UN 2007). The paper listed six potential impacts of climate change 'on issues closely associated with threats to international peace and security': border disputes, migration, energy supplies, other resource shortages, societal stress and humanitarian crises. Emphasis in the concept paper was squarely on ways in which climate change had the potential to exacerbate existing tensions, the least direct way of conceptualizing climate change as a security threat, and the least radical so far as any potential role for the UNSC is concerned. It would seem that, having got agreement to hold the open debate, the United Kingdom was going out of its way not to be provocative. The concept paper pointed out that no other international forum had yet addressed the security implications of a changing climate.

When chairing the day-long meeting, Beckett emphasized that, by holding the debate, the Council was not seeking to preempt the authority of other bodies, including the General Assembly and the ECOSOC (UNSC 2007). A number of developed countries spoke in support of this position. China, the Group of 77 (G77) and the Non-Aligned Movement made the point very clearly however, that they do not deem the UNSC to be the appropriate forum to address climate change. Liu Zhenmin, the Chinese deputy ambassador to the UN, said that the UNSC does not have the 'professional competence' nor is it the 'right decision-making place' for extensive participation leading up to widely acceptable proposals (UNSC 2007, 2). The argument that the Council lacks the necessary expertise to address climate change may be convincing up to a point, but the Council could conceivably access the necessary expertise if required. There is now an abundance of information as to what needs to be done. Alternatively, the Council could work in conjunction with other bodies. Having set up the Counter-Terrorism Committee with inadequate resources, the Council subsequently created a Counter-Terrorism Committee Executive Directorate with about 20 professional staff and presumably something similar could be done in regard to climate change.

Underlying opposition to the Council expanding its interests to deal with climate change seemed to be less a concern with expertise than a fear that Chapter VII powers might be drawn upon to replace the principle of CDR as incorporated into the UNFCCC. The nature of global warming had always meant that a convention to contain the phenomenon could have wide-ranging economic implications: 'a climate convention could constitute a major multilateral economic agreement. The sharing of costs and benefits implied in the convention could significantly alter the economic destinies of individual countries' (Dasgupta 1994, 131). The principle of CDR as incorporated into the UNFCCC embodied what was effectively a deal between developed and developing countries: although all countries in the world would need to cooperate to address climate change, developed countries would take the lead because of their greater historical responsibility for environmental degradation, their ongoing emissions and their economic capacity to take action. Despite accepting as a legal obligation the application of the CDR principle to climate change, leading industrialized countries were reluctant to proceed on this basis and so concessions to developed countries were incorporated into the Kyoto Protocol. These included provisions providing that carbon 'sinks' such as forests could be counted towards meeting a country's obligations and that countries could trade their emissions targets so that the rich could buy credits in order to emit more than their initial allotment (Gardiner 2004, 23). In not ratifying the Kyoto Protocol the US was effectively repudiating the CDR principle it had accepted in the UNFCCC, and more recently developing countries have become concerned that the US wants to substitute the climate change framework established through the UNFCCC with a new framework that does not permit high emitting developing countries to shelter behind the principle of CDR.

The strength of developing country opposition to a conceptualization of climate change as a security threat that might provoke the Council to exercise its Chapter VII powers was underlined by reference to the conceptualization of climate change as a phenomenon whose effects are akin to those of WMD. Kaire Munionganda Mbuenda, representative of Namibia, said that developing countries in particular, had been subjected to what could be described as 'low intensity biological or chemical warfare'. GHGs were destroying plants, animals and human beings. The world knows what caused climate change, and who was responsible. High hopes had been placed in the United Nations Framework Convention on Climate Change and the Kyoto Protocol but developed countries continued to pay lip service to reducing the levels of GHG emissions (UNSC 2007, 16). It was those states that acknowledged climate change as a security threat in the most direct sense that were most receptive to the idea of the Council becoming involved. Robert Aisi, speaking on behalf of the Pacific Islands Forum, emphasized that the very survival of many Pacific Island countries was at risk from rising sea levels and that climate change seemed to be already affecting the intensity and location of tropical cyclones. The impact of climate change on small islands was no less threatening than the dangers guns and bombs posed to large nations. 'The Security Council, charged with protecting human rights and the integrity and security of states, is the paramount international forum available to us', Aisi said (UNSC 2007, 15). Although the Pacific Islands Forum did not expect the Council to get involved in Climate Change Convention negotiations, it did expect the Council to keep the issue under continuous review, to ensure that all countries contributed to solving the problem and that those efforts were commensurate with their resources and capacities.

Conclusions

The international politics of climate change is evolving rapidly. One significant dynamic within this process over the last couple of years has been debate over the question as to whether it is appropriate to label climate change a security threat. This is important not just for rhetorical flourish or factual accuracy but because of the legal and institutional ramifications of doing so. This article has considered the implications for international law of the securitization of climate change and what this might mean in turn for the international politics of climate change. Labelling climate change a threat to international peace and security opens up the possibility of the UNSC taking a lead role in addressing the causes and consequences of climate change, and drawing on its Chapter VII powers to require states to take action on the scale needed to really make a difference. Whereas outcomes of a multilateral negotiating process such as that undertaken on the basis of the UNFCCC are adversely affected by the position of the least willing and rely on the consent and capabilities of the negotiating states for implementation, the UNSC has the authority to dictate to states. And yet ultimately, even UNSC authority depends on the willingness of Council members to reach agreement on the action to be taken, and a perception that what is demanded of states is legitimate. Developing countries have equated the move to securitize climate change as a move away from the principle of CDR and an attempt on the part of the rich to impose policies on developing countries while abdicating responsibility for the climate change crisis.

The UK has been a primary advocate of the securitization of climate change and used its presidency of the UNSC in April 2007 to lead a debate on energy, security and climate. Since then a number of studies have been released pointing to the links between climate change and international security. And yet, at a diplomatic level there has been a reaction against the trend towards securitization of the issue. The General Assembly's informal thematic debate on 'Climate Change as a Global Challenge'—held from 31 July to 2 August 2007—reflected the demand of developing countries that focus be placed on development (United Nations General Assembly 2007). In her opening remarks, the President emphasized that managing climate change was among the greatest challenges currently facing the international community, but emphasis was subsequently placed on the necessity of a sustainable long-term solution to be pursued in the context of the international development agenda and little if any mention was made of 'security' (United Nations General Assembly 2007). A report of the Secretary-General providing an overview of UN activities in relation to climate change that was presented to a two-day conference convened by General Assembly President Srgjan Kerim in February 2008 made multiple references to food security, and one or two to human security, but the claim that climate change will endanger international peace and security was notable for its absence. The G8 Summit Declaration of 7 June 2007 explicitly acknowledged the UN climate process to be the appropriate forum for negotiating future global action on climate change and stressed that further action should be based on the UNFCCC principle of CDR and respective capabilities (G8 Summit 2007, 16). A list of UN system entities to be engaged in future UN climate change activities made no mention of the UNSC (United Nations General Assembly 2008).

Although the diplomatic push to have climate change securitized would at this stage appear to have failed, the inadequacy of all previous and current measures and the uncertainties surrounding negotiation of the post-2012 agreement means that the world as a whole may become increasingly desperate to find a decisive way forward. It is far too early to rule out a constructive role for the UNSC, assisted by a subsidiary body, and acting in tandem with the UNFCCC-Kyoto process. Ironically, however, the practical difference between multilateral negotiating processes and the UNSC legislating is not as great as it might at first seem: neither can succeed until the world's most powerful states demonstrate a preparedness to lead on the issue, make significant changes to their own economies and assist others to meet the challenges ahead in a manner compatible with sustainable development. Australia is one developed country that has undergone a recent change of policy direction on climate change and this has encouraged hopes that the day may soon be approaching when the US follows suit. There are some indications that previous US military opposition to the concept of environmental security is weakening and if the Pentagon does become strongly committed to reducing US reliance on oil this may well generate sufficient momentum to achieve a dramatic shift in attitude on the part of the US as a whole (Sohlman 2008). The rhetoric of climate change securitization cannot be expected to gain a positive reception from the world as a whole until the US leads by example and, when the US does finally do so, the range of possible paths forward will be expanded greatly.

References

ABC News (2007) 'Border security link to climate change: Keelty', 25 September, <http://www.abc.net.au/news/stories/2007/09/25/2042214.htm> , accessed 25 January 2008

Ashton, John (2006) 'World's most wanted: climate change', *BBC News Online Version*, 8 September, <http://news.bbc.co.uk/2/hi/science/nature/5323512.stm> , accessed 29 January 2007

Barnett, Jon (2001) 'Security and climate change', Working Paper 7, Tyndall Centre for Climate Change Research, Norwich, <http://www.tyndall.ac.uk/publications/working_papers/wp7.pdf> , accessed 10 August 2008

Barnett, Jon (2003) 'Security and climate change', *Global Environmental Change*, 13:1, 7–17

Beckett, Margaret (2006) 'Berlin speech on climate and security', Speech given at the British Embassy, Berlin, 24 October, <http://www.britishembassy.de/en/news/items/061024.htm#up> , accessed 22 February 2008

Brock, Lothar (1992) 'Security through defending the environment: an illusion?' in Elise Boulding (ed) *New agendas for peace research: conflict and security reexamined* (Boulder, Colorado: Lynne Rienner), 79–102

Brock, Lothar (1997) 'The environment and security: conceptual and theoretical issues' in Nils Petter Gleditsch (ed) *Conflict and the environment* (Dordrecht, The Netherlands: Kluwer), 17–34

Brown, Lester (1977) *Worldwatch Paper 14: Redefining national security*, (Washington DC: Worldwatch)

Buckland, Ben (2007) 'A climate of war? Stopping the securitization of global climate change', Paper written for the International Peace Bureau, Geneva, Switzerland

Busby, Joshua W (2007) 'Climate change and national security: an agenda for action', CSR 32 (New York: Council on Foreign Relations), < http://www.cfr.org/content/publications/attachments/ClimateChange_CSR32.pdf> , accessed 10 August 2008

Buzan, Barry, Ole Waever and Jaap de Wilde (1998) *Security: a new framework for analysis* (Boulder, Colorado: Lynne Rienner)

Chatham House (2007) 'Military chief: climate change and instability like pouring petrol on fire', Media release, 25 June, <http://www.chathamhouse.org.uk/news/view/-/id/382/> , accessed 26 February 2008

Clark, Andrew (2007) 'Climate change threatens security, UK tells UN', *The Guardian*, 18 April, <http://www.guardian.co.uk/environment/2007/apr/18/greenpolitics.climatechange/print> , accessed 10 August 2008

CNA Corporation (2007) *National security and the threat of climate change*, < http://securityandclimate.cna.org/> , accessed 8 February 2008

Dalby, Simon (1997) 'Contesting an essential concept: reading the dilemmas in contemporary security discourse' in Keith Krause and Michael C Williams (eds) *Critical security studies: concepts and cases* (Minneapolis: University of Minnesota Press), 3–31

Dasgupta, Chandrashekhar (1994) 'The climate change negotiations' in Irving M Mintzer and J Amber Leonard (eds) *Negotiating climate change: the inside story of the Rio Convention* (Cambridge: Cambridge University Press), 129–148

Deudney, Daniel (1990) 'The case against linking environmental degradation and security', *Millennium: Journal of International Studies*, 19:3, 461–476

De Wet, Erika (2004) *The Chapter VII powers of the United Nations Security Council* (Oxford: Hart Publishing)

Dunlop, Ian (2007) 'Climate change is a war that we must fight', *The Age*, 23 October

The Economist (2007) 'Drying up and flooding out', 10 May, <http://www.economist.com/world/africa/displaystory.cfm?story_id = 9163426>, accessed 27 February 2008

Egelko, Bob (2006) 'Climate change is as serious as WMD: Annan', *Agence France Presse*, 15 November, <http://www.commondreams.org/headlines06/1115-02.htm>, accessed 20 February 2008

Falk, Richard (1971) *This endangered planet: prospects and proposals for human survival* (New York: Random House)

Flannery, Tim (2007) 'The long view', *The Age*, 31 October, <http://www.theage.com.au/>, accessed 20 February 2008

G8 Summit 2007 (2007) 'Growth and responsibility in the world economy', Summit Declaration, 7 June, <http://www.g-8.de/Content/EN/Artikel/__g8-summit/anlagen/2007-06-07-gipfeldokument-wirtschaft-eng,property = publicationFile.pdf>, accessed 10 August 2008

Gardiner, Stephen (2004) 'The global warming tragedy and the dangerous illusion of the Kyoto Protocol', *Ethics and International Affairs*, 18:1, 23–39

Gow, David (2008) 'Prince of Wales likens climate change battle to war', *The Guardian*, 14 February, <http://www.guardian.co.uk/environment/2008/feb/14>, accessed 20 February 2008

Homer-Dixon, Thomas (2001) *Environment, scarcity, and violence* (Princeton: Princeton University Press)

Houghton, John (2003) 'Global warming is now a weapon of mass destruction', *The Guardian*, 28 July, <http://politics.guardian.co.uk/green/comment/0,9236,1007302,00.html>, accessed 10 August 2008

Howden, Daniel (2007) 'Bali conference: diplomats warned that climate change is a security issue, not a green dilemma', *The Independent*, 6 December, <http://www.independent.co.uk/environment/climate-change/bali-confe ... ge-is-security-issue-not-a-green-dilemma-763247.html>, accessed 25 January 2008

Intergovernmental Panel on Climate Change (2007) *Climate change 2007: Mitigation of climate change. Contribution of Working Group III to the fourth assessment report of the Intergovernmental Panel on Climate Change* (Cambridge: Cambridge University Press)

Kellow, Aynsley (2008) 'Lessons not learned in environmental governance: international climate policy beyond Kyoto', *Asia Pacific Journal of Environmental Law*, 11:1–2, 101–120

Levey, Marc A (1995) 'Is the environment a national security issue?', *International Security*, 10:2, 35–62

Luck, Edward C (2006) *UN Security Council: practice and promise* (London: Routledge)

Matheson, Michael J (2006) *Council unbound: the growth of UN decision making on conflict and post-conflict issues after the Cold War* (Washington, DC: United States Institute of Peace Press)

Mathews, Jessica Tuchman (1989) 'Redefining security', *Foreign Affairs*, 68:2, 162–177

Monbiot, George (2006) 'Drastic action on climate change is needed now—and here's the plan', *The Guardian*, 31 October, <http://www.guardian.co.uk/commentisfree/2006/oct/31/economy.politics>, accessed 20 February 2008

Murphy, Michael K (1998–1999) 'Achieving economic security with swords as ploughshares: the modern use of force to combat environmental degradation', *Virginia Journal of International Law*, 39:4, 1181–1220

Nicoll, Alexander (2007) *Strategic survey 2007: the annual review of world affairs* (Oxford: Oxford University Press)

Nordas, Ragnhild and Nils Petter Gleditsch (2007) 'Climate change and conflict', *Political Geography*, 26:6, 627–638

Oxfam International (2007) 'Oxfam analysis of the Bali conference outcomes', 15 December, <http://www.oxfam.org.uk/resources/policy/climate_change/bali_outcomes.html>, accessed 7 March 2008

Paskal, Cleo (2007) 'How climate change is pushing the boundaries of security and foreign policy', Chatham House Briefing Paper, Energy, Environment and Development Programme, EEDP CC BP 07/01, June, <http://www.chathamhouse.org.uk/>, accessed 21 February 2008

Penny, Christopher K (2007) 'Greening the Security Council: climate change as an emerging "threat to international peace and security"', *International Environmental Agreements: Politics, Law and Economics*, 7:1, 35–71

Podesta, John and Peter Ogden (2007–2008) 'The security implications of climate change', *Washington Quarterly*, 31, 115–138

Royal Society (2005) 'Climate change effects can be compared to "weapons of mass destruction" warns Lord May', Press release, 28 November, <http://royalsociety.org/news.asp?id = 3923>, accessed 10 August 2008

Scott, Shirley V (2004) *The political interpretation of multilateral treaties* (Leiden, The Netherlands: Martinus Nijhoff)

Smith, Dan and Janani Vivekananda (2007) *A climate of conflict: the links between climate change, peace and war* (London: International Alert)

Sohlman, Eva (2008) 'Flight of the green hawks: why the Pentagon is on a green mission', *Washington Life Magazine*, April, <http://washingtonlife.com/issues/april-2008/special_report/index.php>, accessed 17 July 2008

Szasz, Paul C (2002) 'The Security Council starts legislating', *American Journal of International Law*, 96:4, 901–905

UK Mission to the UN, New York (2007) 'Energy, security and climate', UK Concept Paper, 23 April, <http://www.britishembassy.gov.uk/servlet/Front?pagename = OpenMarket/Xcelerate/ShowPage&c = Page&cid = 1183543073739>, accessed 4 February 2008

United Nations Environment Programme (2007) *Sudan: post-conflict environmental assessment* (Nairobi, Kenya: UNEP)

United Nations General Assembly (2007) 'Informal thematic debate: "Climate change as a global challenge"', Informal summary of the debate, <http://www.un.org/ga/president/61/follow-up/thematic-climate.shtml>, accessed 5 February 2008

United Nations General Assembly (2008) 'Overview of United Nations activities in relation to climate change: Report of the Secretary-General', A/62/644, 10 January

United Nations Security Council (2007) 'Security Council holds first-ever debate on impact of climate change', SC/9000, 17 April

World Commission on Environment and Development (1987) *Our common future* (Oxford: Oxford University Press)

WorldPublicOpinion.Org (2007) 'International polling on climate change', 6 December, <http://www.worldpublicopinion.org/>, accessed 5 February 2008

Inequality and the global climate regime: breaking the north-south impasse

Bradley C Parks and J Timmons Roberts[1]

Introduction

Since the early 1990s, virtually all developing countries have refused to adopt greenhouse gas emission reduction commitments in the name of *fairness*. In fact, the very suggestion that poorer nations limit their industrial growth has led to a hostile negotiating environment. At the Kyoto summit, China's lead negotiator said, 'In the developed world only two people ride in a car, and yet you want us to give up riding on a bus.' Brazil's Chancellor Luiz Felipe Lampreia said flatly 'We cannot accept limitations that interfere with our economic development' (Rossi 1997, A-15). Ten years later, after rounds and rounds of painstaking international negotiations, the position of most developing countries has not changed significantly. The head of China's National Development and Reform Commission stated in mid-2007 that '[i]t is neither fair nor acceptable to us to impose too early, too abruptly, or too bluntly measures which one would ask of developed countries. ... People are not putting the blame on those countries with large historical emissions [and] high per capita emissions' (Ford 2007).

---

[1] The views expressed in this article are the authors' own and do not necessarily represent the views of the Millennium Challenge Corporation.

We argue that the stalemate in north-south climate negotiations is unlikely to be resolved in the absence of aggressive efforts to address issues of inequality and justice. Inequality and justice have been central issues at every major environmental conference since the 1972 UN Conference on the Human Environment in Stockholm, Sweden: Nairobi in 1982, Rio in 1992, Rio+5 in New York and Johannesburg in 2002. At negotiations leading up to the 1992 Earth Summit, southern countries feared limits on their efforts to grow economically and care for the basic needs of their people, but powerful industrialized countries such as the United States (US) refused to curtail their own excesses unless poor nations did the same. In the context of climate change negotiations, all states came under intense pressure to 'do something', and 132 countries eventually did sign the United Nations Framework Convention on Climate Change (UNFCCC). But the treaty avoided tough details. It called on nations to 'protect the climate system . . . on the basis of equity and in accordance with their common but differentiated responsibilities and respective capabilities' (UNFCCC 1992, articles 3 and 4). However, consensus around this general principle did not translate into broad-based agreement on the issue of actual obligations (Baumert 2002).

Before the ink had even dried on the UNFCCC agreement, rich countries began to back away from earlier promises of technology transfer and technical assistance to the developing world (Najam 2002). Future rounds of negotiations foundered on the issues of equity and justice—at the first Conference of the Parties (COP) in Berlin in 1995 and subsequent meetings in Geneva, Kyoto, The Hague, Bonn, Marrakech, Buenos Aires, Montreal, Nairobi and Bali. President Bill Clinton agreed to sign on to the Kyoto Protocol in 1997, but even before he did, the US Senate voted 95 to 0 to support the Byrd-Hagel Resolution, which would block any 'unfair' treaty that did not require poor nations to also limit carbon dioxide emissions.

The Byrd-Hagel Resolution was widely denounced by leaders in the developing world. Southern policy-makers and activists were quick to point out that the average US citizen dumps as much greenhouse gas into the atmosphere as five Chinese or seventeen Indians, and that developing countries are immeasurably more vulnerable to rising tides, tropical storms, droughts and flooding than rich nations (Roberts and Parks 2007; Agarwal et al 2001). Policy-makers and 'opinion leaders' in the developing world suggested that the US was coopting and thereby undermining their position of 'climate injustice'.

Why does inequality matter?

The absence of an effective global agreement that brings rich and poor nations together to protect the climate raises broad questions about the determinants of interstate cooperation related to transnational environmental threats. Over the past 20 years, scholars have argued that outcomes in international environmental politics are shaped by material self-interest, bargaining power and the ability to strong-arm weaker states through more coercive forms of power (Sprinz and Vaahtoranta 1994; Victor 2001). Others have emphasized the importance of exogenous shocks and crises, salient solutions, a scientific burden of proof, environmental nongovernmental organizations (NGOs), postmaterialist values, epistemic communities, transnational activist networks, corporate nonstate actors, intergovernmental organizations and political leadership (Young 1994;

Wapner 1995; Haas 1990; Keck and Sikkink 1998; Levy and Kolk 2002; Meyer et al 1997). Yet interestingly, one of the variables often singled out by southern policy-makers as a major impediment to cooperation—global inequality—has not received much scholarly attention.[2]

In this article, we identify specific causal channels through which inequality can influence the prospects for international cooperation on global climate change. Inequality, we argue, can dampen utility-enhancing cooperative efforts by reinforcing 'structuralist' world-views and causal beliefs, polarizing policy preferences, making it difficult to coalesce around a socially shared understanding of what is 'fair', eroding conditions of trust, generating divergent and unstable expectations about future behaviour, and creating incentives for zero-sum and negative-sum behaviour.

There are three broad types of inequality that we believe figure prominently in climate change negotiations: climate-related inequality, inequality in international environmental politics and inequality in international economic regimes. First, we identify the significant asymmetries in which countries will suffer most and earliest from climate change, which countries are most responsible for the problem and which countries are expected to shoulder most of the atmospheric clean-up. We also find a significant amount of 'bleed over' from other issue areas, such as trade, investment, debt, intellectual property rights, biodiversity and desertification. In many of these international regimes, developing countries feel as though their concerns regarding fair processes and outcomes have been marginalized. We briefly describe each of these inequalities and explain how their existence, and the industrialized world's reaction to them, has made it more difficult for rich and poor nations to forge a post-2012 global climate pact. We conclude by providing several historical examples that illustrate how countries with highly disparate world-views, causal beliefs, principled beliefs and policy positions have resolved their differences and cooperated on issues of mutual interest.

Inequality in the greenhouse

Inequality in responsibility for climate change

A casual observer might think that the best way to resolve the issue of responsibility for climate change would be to give all humans equal atmospheric rights and assign responsibility to individuals based on how much 'environmental space' they use. This is a basic rule of civil justice and kindergarten ethics: those who created a mess should be responsible for cleaning up their fair share. But in international politics things are not so simple.

With only four per cent of the world's population, the US is responsible for over 20 per cent of all global emissions. That can be compared to 136 developing countries that together are only responsible for 24 per cent of global emissions (Roberts and Parks 2007). Poor countries therefore remain far behind wealthy countries in terms of emissions per person. Overall, the richest 20 per cent of the world's population is responsible for over 60 per cent of its current emissions of

[2] There are, of course, a few noteworthy exceptions (see Chasek et al 2006; Najam 2004; Müller 1999).

greenhouse gasses. That figure surpasses 80 per cent if past contributions to the problem are considered, and they probably should be, since carbon dioxide (the main contributor to the greenhouse effect) remains in the atmosphere for over one hundred years.

These vast disparities have shaped different proposals for cleaning up the atmosphere. The Kyoto Protocol, as it was negotiated in 1997, was based on *grandfathering*: the notion that countries should reduce their emissions incrementally from a baseline year (1990). Large polluters, therefore, had their high discharges of greenhouse gases grandfathered in, and committed to relatively minor emission reductions averaging 5.2 per cent for the foreseeable future. The *carbon intensity* approach, introduced by the World Resources Institute and favoured by the second Bush administration starting in 2002, calls for voluntary efficiency changes to drive emission reductions. Under this approach, the goal is to have strong economic growth with as few carbon emissions as possible (Baumert 2002). Both of these proposals have the modest effect of departing incrementally from the current status quo without radical requirements on powerful countries.

On the other side of the spectrum are two proposals that strongly favour developing countries: *historical responsibility* and per capita *contraction and convergence*. India, China and much of the developing world favour a per capita approach, in which each person on earth is given an equal right to the ability of the atmosphere to absorb carbon. Under the per capita proposal, countries whose per capita consumption of fossil fuels is significantly lower than the world average would be given significant room to grow and emit. Most per capita plans would allow them to trade their extra carbon emission credits for the capital they need for development. By comparison, countries with highly fossil-energy-intensive economies would face sharp requirements to cut their consumption of fuels. Brazil also introduced a proposal in 1997 that would take into account the amount of damage done by nations in the past to the atmosphere's ability to absorb more greenhouse gases (Baumert 2002). This *historical responsibility* approach places the onus on countries that put greenhouse gases in the atmosphere in past decades to reduce their emissions quickly, most notably the United Kingdom (UK) and the US: a few developing countries have supported this approach and demanded that some indemnification be paid for the so-called 'carbon debt'.

Importantly, each of these four carbon accounting approaches—grandfathering, carbon intensity, historical responsibility and per capita contraction and convergence—represents a different social understanding of fairness. Grandfathering represents the entitlement principle that individuals are entitled to what they have produced. Carbon intensity represents the utilitarian principle that since everyone is worse off in the absence of joint gains, inefficient solutions are also unjust. Historical responsibility represents 'the polluter pays' principle. Finally, the per capita approach represents the egalitarian principle that every human should have equal rights to global public goods, such as atmospheric stability. These different perceptions of fairness are shaped to a large extent by the highly disparate positions that countries occupy in the global hierarchy of economic and political power. In this way, we argue that inequality has a dampening effect on cooperation by polarizing policy preferences and making it difficult for countries to arrive at a socially shared understanding of what is 'fair'.

Inequality in vulnerability to climate change

The scientific community agrees that carbon emissions will create a warmer and wetter atmosphere, and, in turn, increase flooding, hurricanes, forest fires, winter storms, and drought in arid and semi-arid regions. Climatologists have observed a sharp upswing in the frequency, magnitude and intensity of hydro-meteorological disasters over the past two decades—the five warmest years on historical record were 1998, 2002, 2003, 2005 and 2007—and hydro-meteorological disasters have more than doubled since 1996 (Goddard Institute for Space Studies 2008).

Whereas climate change is often described as 'everybody's problem' or a 'global public bad', hydro-meteorological impacts are distributed socially across human populations (Kaul et al 1999). Some countries and communities will suffer most and earliest, and generally they are not those that caused the problem. According to the latest predictions of the Intergovernmental Panel on Climate Change (IPCC), rapidly expanding populations in Africa, Asia and Latin America are suffering disproportionately from more frequent and dangerous droughts, floods and storms (IPCC 2007). The World Bank reports that '[b]etween 1990 and 1998, 94 per cent of the world's disasters and 97 per cent of all natural-disaster-related deaths occurred in developing countries' (Mathur et al 2004, 6).

Table 1 shows the top five climate-related disasters by type and human impact. It demonstrates very clearly that although natural disasters may seem to have an underlying random origin, their impacts are anything but random. The same countries are consistently among those who suffer the most from climate disasters. Poor countries in Asia, Africa and Central America repeatedly top the list of countries with the most deaths and livelihood disruptions. The totals over 20 years are sobering, with 62 million Bangladeshis and nearly 50 million Chinese made homeless by climate disasters. Over one million were made homeless in Laos, Sudan, India, Sri Lanka, Vietnam, Philippines and Pakistan. However, many of these countries are very large in population size, which makes it difficult to judge the severity of impact for a given country. Table 2 adjusts the total number of climate-related fatalities and cases of homelessness by population size in order to provide more comparable cross-country estimates of hydro-meteorological risk. The results are startling. In relative terms, ten times more, and in some cases hundreds of times more, people are dying in the developing world than in the US and the UK. Less than one-seventh of one per cent of the US population was made homeless by hydro-meteorological disasters between 1980 and 2002.[3] By comparison, in Bangladesh 45 per cent of the population was at some point made homeless during the same period.

Inequalities in vulnerability to climate change are important because poor countries suffering from rising sea levels, devastating droughts and storms, lower

[3] Despite the fact that Hurricane Katrina was one of the five deadliest hurricanes ever experienced by the US, the overall human impact at the national level was significantly lower than in many developing countries that have faced storms of similar strength. According to the Emergency Events database (EM-DAT), approximately 1,833 people, or 0.000006% of the total US population, died in Hurricane Katrina.

Table 1. Top five climate disasters by type and human impact, 1980–2002, as measured by number of people killed and made homeless*

Top five wind storms by number of human deaths

Country	*Year*	*Number killed*
Bangladesh	1991	138,866
Honduras	1998	14,600
Bangladesh	1995	10,000
India	1999	9,843
Philippines	1991	5,956

Top five floods by number of human deaths

Country	*Year*	*Number killed*
Venezuela	1999	30,000
China	1980	6,200
China	1998	3,656
China	1996	2,775
Bangladesh	1987	2,379

Top five droughts by number of human deaths

Country	*Year*	*Number killed*
Ethiopia	1984	300,000
Sudan	1984	150,000
Mozambique	1985	100,000
Chad	1984	3,000
China	1988	1,400

Top five heat waves by number of human deaths

Country	*Year*	*Number killed*
India	1998	2,541
India	2002	1,030
Greece	1987	1,000
United States	1995	670
India	1995	558

Top five wind storms by number of people homeless

Country	*Year*	*Number killed*
China	1998	11,000,000
Bangladesh	1988	2,000,000
Philippines	1994	1,170,875
Philippines	1990	1,110,020
Philippines	1991	1,048,024

Top five floods in number of people homeless

Country	*Year*	*Number killed*
Bangladesh	1987	28,000,000
Bangladesh	1988	28,000,000
China	1998	15,850,000
China	1994	5,600,000
Pakistan	1992	4,698,670

*Rankings are authors' calculations based on reported figures as assembled by OFDA CRED/EM DAT 2004 (Center for Research on the Epidemiology of Disasters/ The International Emergency Disasters Database) 2008. (Brussels: CRED), <http://www.emdat.be/>.

agricultural yields and increased disease burdens are unlikely to be enthusiastic about cleaning up an environmental problem that the industrialized world created in the first place.[4] Indeed, stark inequalities in vulnerability have poisoned the negotiating atmosphere. 'If climate change makes our country uninhabitable', said Bangladeshi Atiq Rahman during the 1995 Berlin negotiations, 'we will march with our wet feet into your living rooms' (Athanasiou and Baer 2002, 23). At every subsequent COP, developing countries have underscored their small contribution to

[4] Conversely, one might argue that self-interest would make more vulnerable countries more likely to join global efforts to reduce greenhouse gas emissions (see Sprinz and Vaahtoranta 1994).

Table 2. Top 20 countries by number of people killed or made homeless between 1980 and 2002, adjusted by population size

Rank	*Country*	*Killed per thousand population*	*Rank*	*Country*	*Per cent of population made homeless*
1	Mozambique	5.55	1	Tonga	50.51
2	Sudan	4.84	2	Bangladesh	45.51
3	Ethiopia	4.78	3	Laos	18.94
4	Honduras	2.40	4	Samoa	17.61
5	Bangladesh	1.23	5	Sri Lanka	13.99
6	Nicaragua	0.73	6	Solomon Islands	13.67
7	Swaziland	0.60	7	Marshall Islands	11.76
8	Papua New Guinea	0.49	8	Antigua and Barbuda	11.55
9	Vanuatu	0.49	9	Philippines	10.34
10	Micronesia, Fed. States	0.42	10	Guam	10.07
11	Chad	0.41	11	Virgin Islands (US)	9.92
12	St Lucia	0.39	12	Maldives	8.20
13	American Samoa	0.37	13	Comoros	7.08
14	Somalia	0.36	14	St. Lucia	6.99
15	Djibouti	0.28	15	Pakistan	6.14
16	Philippines	0.27	16	Vanuatu	5.53
17	Haiti	0.25	17	Somalia	5.47
18	Tajikistan	0.24	18	Vietnam	4.76
19	Solomon Islands	0.23	19	Benin	4.56
20	Belize	0.19	20	Sudan	4.04
	United States	0.03		United States	0.14

the problem of climate change and their extreme vulnerability to its impacts (Müller 2001).[5] Some climate policy analysts dismiss this line of argumentation as mere posturing, but a 2008 European Union report warns that '[c]limate change impacts will fuel the politics of resentment between those most responsible for climate change and those most affected by it' (European Union 2008, 5).

Inequality in (expected) clean-up

There are also inequalities in who is currently doing something to reduce greenhouse gas emissions and which countries will likely bear the greatest burden of atmospheric clean-up in the future. Here again, the disparities are very significant. The first important point is that, while northern governments are trying to convince the southern governments that they need to rein in their greenhouse gas emissions, most of them are not doing so in their own countries. Under the Kyoto Protocol, 'Annex I' (developed) countries committed to a 5.2 per cent (average) reduction in greenhouse gas emissions (below 1990 levels) by 2012. However, Table 3 shows that, with the exception of several European countries,

[5] Young notes that '[s]ome northerners may doubt the credibility of [threats from southern nations to damage the global climate] and advocate a bargaining strategy that offers few concessions to the developing countries. But such a strategy is exceedingly risky. Many of those located in developing countries are increasingly angry and desperate. ... Faced with this prospect, northerners will ignore the demands of the South regarding climate change at their peril' (1994, 50).

Table 3. Change in GHG emissions (1990–2004, excluding the effects of land use, land-use change and forestry)

Country	*Per cent change in GHG emissions*
Australia	25.10
Austria	15.70
Belgium	1.40
Canada	26.60
Denmark	−1.10
Finland	14.50
France	−0.80
Germany	17.20
Ireland	23.10
Italy	12.10
Japan	6.50
Netherlands	2.40
New Zealand	21.30
Norway	10.30
Portugal	41.00
Spain	49.00
Sweden	−3.50
Switzerland	0.40
United Kingdom	−14.30
United States	15.80
Annex I Parties to Kyoto Protocol	2.90

greenhouse gas emissions have risen significantly throughout the industrialized world since 1990. As a result, the so-called 'demandeurs' of global climate protection face a serious credibility problem. According to Baumert and Kete (2002, 6), '[m]any developing countries believe that the industrialized countries lack credibility on the issue of international cooperation to curb greenhouse gas emissions, having done little to address a problem largely of their own making.'

Many industrialized countries have also decided that rather than making cuts at home, they would prefer to achieve their emission reduction commitments by funding activities in developing countries. From a cost efficiency perspective, this makes eminent sense: the greatest opportunities for low-cost emissions reductions exist in the developing world (Stavins and Olmstead 2006). However, as we describe in greater detail below, the last 35 years of global environmental negotiations have demonstrated that developing countries have deeply held distributional concerns, which can be a significant impediment to international cooperation. According to one Group of 77 (G77) expert, the south's 'principal fear ... [is] that the North is using environmental issues as an excuse to pull up the development ladder behind it—[a suspicion which] has remained unallayed through two decades of environmental diplomacy' (Najam 1995, 249). Joanna Depledge (2002), a former UNFCCC Secretariat staff member, has similarly reported that many 'Annex II' countries fear efforts to curb carbon emissions in the developing world will effectively place a 'cap' on their economic growth.

It is also important to note that even in those northern countries which appear to have reduced or stabilized their greenhouse gas emissions, there are good reasons to believe that far less has actually been achieved than the summary statistics suggest. New research shows that many 'service-exporting' OECD countries, which specialize in areas like banking, tourism, advertising, sales, product design, procurement and distribution, are actually 'net-importers' of carbon-intensive goods coming primarily from developing countries. They therefore emit no less; they simply displace their emissions (Machado et al 2001; Muradian et al 2002; Heil and Selden 2001).[6]

Finally, there are expected inequalities in the atmospheric clean-up obligations that will be negotiated in the post-Kyoto era. In order to understand this, it is helpful to revisit the basic science. Left unattended, greenhouse gas emissions are expected to rapidly accumulate in the earth's atmosphere over the next century and global average temperatures are expected to rise by 1.4–5.8°C (IPCC 2007). Between 1996 and 2004, total global carbon emissions increased from approximately six billion tons of carbon equivalent (GtC) per year to seven billion GtC. Yet scientists warn that to avoid 'dangerous anthropogenic interference with the climate system', we must cap atmospheric CO_2 concentrations somewhere between 450 and 550 parts per million (ppm), or at roughly 9.4 billion GtC per year (Baumert and Kete 2002). Very substantial emissions reductions will therefore be necessary in the near term to stabilize the climate. According to current trends, the world will likely reach 9.8 billion GtC by 2020.

The fundamental political question is who will be responsible for the bulk of *future* global greenhouse gas emissions reductions. The current accumulated stock of CO_2 in the atmosphere is largely the responsibility of rich, industrialized countries. However, growth in future emissions is expected to primarily take place in the developing world. Developed countries are on track to register roughly 1 per cent annual economic growth, whereas developing countries are averaging 3.5 per cent growth and are expected to maintain a similar growth trajectory in the coming decades. At the same time, the global population will continue to expand—from 6 billion to 8.4–12 billion people over the next 100 years—and the lion's share of this growth will take place in the developing world. Together economic development and population growth are expected to increase non-Annex I carbon emissions from 31 per cent of the total in 1990 to 60 per cent in 2030 (Wheeler and Ummel 2007; Blanchard et al 2003). Climate stabilization therefore demands that 'the South... accept the necessity of serious, costly mitigation, and immediately embark on a low-carbon development path' (Wheeler and Ummel 2007, 10).

Inequality in international environmental regimes

Climate negotiations are deeply embedded in the broader context of north-south environmental relations. In 1972, at the first international conference on the environment in Stockholm, Sweden, it quickly became evident that no consensus would emerge between developed and developing countries on the issue of global environmental protection. 'Late-developers' feared restrictions on their economic growth, emphasized the north's profligate use of planetary resources, and pushed

[6] This pattern has been celebrated as a clear indication that rich countries are becoming more 'postmaterialist', or 'postindustrial'. However, these service economies still require extraordinary levels of energy and materials. New research has demonstrated that the production of the material goods (and their effluents) has shifted over time to poorer countries. As such, the material-intensive imports required by rich countries have carbon emissions 'embodied' within them (Machado et al 2001).

for a redistributive programme that would benefit them economically and hasten the transition towards industrialization. Developed countries wanted northern consumption off the negotiating table, southern population growth on the agenda, and nonbinding language on issues of financial assistance and technology transfer. Neither negotiating bloc was willing to budge, and deeply held feelings of marginalization and injustice among poor nations made for an adversarial negotiating atmosphere (Haas et al 1993).

The south's confrontational approach intensified in the late 1970s under the banner of the 'New International Economic Order' (NIEO). During this period, developing countries put forth a 'series of proposals . . . which included significant wealth redistribution, greater LDC participation in the world economy, and greater Third World control over global institutions and resources' (Sebenius 1991, 128). Although several significant successes were achieved (Ruggie 1983), many of the hardest-fought victories were rolled back during the years of Thatcherism, Reaganism and neoliberalism. During this period, late developers also became strident in their criticism of northern environmentalism—an environmentalism which they perceived as 'pull[ing] up the development ladder' (Najam 1995).

In future rounds of negotiations, on issues such as biodiversity, desertification and climate change, there were calls for increased financial compensation and more equitable representation (Sell 1996). Debate over the voting structure of the Global Environmental Facility, which distributes hundreds of millions of dollars of environmental aid each year, became especially conflict-ridden. Poor and middle-income countries protested 'donor dominance' and the lack of transparency in decision-making, whereas rich, industrialized countries insisted that only the 'incremental costs' of global environmental projects be financed (Keohane and Levy 1996). North-south environmental relations also suffered several important setbacks after the 1992 Rio Earth Summit (Raustiala 1997). Developed countries agreed to underwrite the participation of less developed countries in any global environmental accord to come out of Rio. Specifically, they agreed to a financial package of 100 billion US dollars (USD) a year in new and additional concessionary funds for 'sustainable development' and 15 billion USD for global environmental issues (Robinson 1992). However, wealthy OECD countries failed to honour their policy commitments (Najam 2002). At the end of the 1990s, approximately three billion USD a year was being allocated for global environmental issues (20 per cent of the Rio promise), and approximately seven billion USD a year was being given for local environmental projects (Hicks et al 2008). The reasons varied: recessions at home, new electoral coalitions in power, executive commitments that legislatures refused to ratify or sustain, or simple backsliding.

Then in 1997, at the UN General Assembly Special Session for Review and Appraisal of Agenda 21 (UNGASS), developing countries sought to strengthen the 'sustainable development' agenda by linking the issues of climate change, forests and biodiversity to issues of trade, investment, finance and intellectual property rights. This was flatly rejected by rich nations (Sandbrook 1997). Three years later, at the COP6 climate negotiations, developing country delegations expressed outrage after the (western) chairs allegedly deleted text that had been agreed upon earlier (Dessai 2001). The G77 and China also charged that many of the important decisions affecting developing countries were being made in nontransparent 'Green Room' meetings, attended only by powerful countries. This set the stage for the 2002 World Summit on Sustainable Development

(WSSD), where one reporter noted that 'effective governance is not possible under the prevailing conditions of deep distrust' (Najam 2003, 370).[7] As we argue at greater length below, this lack of trust in north-south relations has proven to be a major obstacle to the creation of a post-2012 global climate pact.

Inequality in international economic regimes

International climate negotiations are also situated within the broader context of north-south economic relations. Stephen Krasner once said that there are 'makers, breakers, and takers' in international relations, and there is little question that developing countries are generally 'takers' in international economic regimes (Krasner 1978). '[T]he "price" of multilateral rules', explains Shadlen, 'is that [Least Developed Countries–LDCs] must accept rules written by—and usually for—the more developed countries' (Shadlen 2004, 6). Gruber (2000) argues that powerful states—particularly those with large markets—possess 'go-it-alone power' in that they can unilaterally eliminate the previous status quo and proceed gainfully with or without the participation of weaker parties. He adds,

> Faced with a choice between joining the winners' new cooperative system or being completely shut out, the losers enthusiastically submit their applications for membership. They do so, however, only because the winners' actions have had the effect of removing the status quo from their choice sets, leaving them with what they view as a bad option (cooperating with the winners) and an even worse alternative (incurring the costs of exclusion). (Gruber 2000, 8)

Wade refers to a so-called 'shrinking of development space', and argues that 'the rules being written into multilateral and bilateral agreements actively prevent developing countries from pursuing the kinds of industrial and technology policies adopted by the newly developed countries of East Asia and by the older developed countries when they were developing' (2003, 622). Similarly, Birdsall et al explain how the callous—and at times opportunistic—actions of western governments have made upward mobility in the international division of labour difficult:

> In the context of international trade agreements in particular, developing countries have been asked to take on obligations that have been clearly inimical to their development interests. Perhaps the most egregious example of this in recent times has been the WTO's intellectual property agreement, TRIPs [The Agreement on Trade-Related Aspects of Intellectual Property Rights]. TRIPs will have the effect on poor countries of increasing the costs of and reducing access to essential medicines and this at a time when one of the worst health epidemics ever known by man—AIDS—ravages the developing world. The flip side of the costs to these countries is the profits that will be transferred from consumers and taxpayers in poor countries to pharmaceutical companies in the rich world. In other words, TRIPs will entail a pure transfer of rents from poor to rich. (2005, 8)[8]

Other scholars of international political economy have highlighted the fact that the governance structures of international financial institutions, like the International

---

[7] Another observer reported, 'the reservoir of mistrust and cynicism between prosperous Northern nations and poor Southern ones has widened since the last Earth Summit in Rio de Janeiro a decade ago' (Jeter 2002).

[8] Also see Rodrik (2001).

Monetary Fund and World Bank, prevent the institutions' main clients (developing countries) from having any significant voting power (Woods 1999; Wade 2003).

These inequalities of opportunity have had a significant impact on how developing countries approach global environmental negotiations. Porter and Brown note that 'developing states' perceptions of the global economic structure as inequitable has long been a factor in their policy responses to global environmental issues' (1991, 124; see also Chasek et al 2006). As we argue in greater detail below, when powerful states disregard weaker states' position in the international division of labour in areas where they possess structural power (as in international economic regimes), they run a high risk of weaker states 'reciprocating' in policy areas where they possess more bargaining leverage (as in international environmental regimes).[9]

How global inequality influences international climate negotiations

We now explore the specific causal mechanisms through which inequality—in opportunity, political power and distributional outcomes—influences global climate negotiations. We argue that global inequality makes it more difficult for rich and poor nations to identify socially shared understandings of 'fair' solutions. And even when rich and poor countries can agree on general fairness principles, the heterogeneity in preferences generated by global inequality aggravates disagreements about how to make those principles operational. Global inequality also contributes to conditions of generalized mistrust, which in turn makes developing countries—countries that possess strong preferences for 'cheap' economic development and weak preferences for stringent environmental policies—more inclined to pursue self-damaging policies. Southern suspicion of northern behaviour and its inability to constrain northern opportunism also promotes risk-averse behaviour and defensive negotiating strategies.

Structuralist world-views and causal beliefs

One of the most important pathways through which global inequality can impede cooperation is by promoting 'structuralist' world-views and causal beliefs. Goldstein and Keohane define world-views as ideas that 'define the universe of possibilities for action' (1993, 9). For example, culture, religion, rationality, emotion, ethnicity, race, class, gender and identity all shape the way that humans (including policy-makers) perceive the opportunities and challenges facing them. As such, having a world-view implies '[limited] choice because it logically excludes other interpretations of reality, or at least suggests that such interpretations are not worthy of sustained exploration' (12). By limiting one's menu of available options, world-views and *causal beliefs*[10] have an instrumental impact on how cost-benefit calculations are conducted. They also influence the very way in which actors come up with their own policy agendas. For example, depending on one's position in the

[9] Gupta rightly notes that '[developing country] negotiators tend to see issues holistically and link the issue to all other international issues. Thus linkages are made to international debt, trade and other environmental issues such as desertification' (2000, 58).

[10] Causal beliefs are 'beliefs about cause-effect relationships which derive authority from the shared consensus of recognized elites' (Goldstein and Keohane 1993, 9–10).

international system, states may seek to maximize absolute gains, relative gains, social (fairness) preferences or emotional utility. Highly risk-averse governments may want to freeze the status quo (Shadlen 2004; Gruber 2000; Abbott and Snidal 2000). Leaders who feel cheated by others may seek to punish their enemies or strengthen their relative power, regardless of the efficiency implications.[11] Those who see themselves as marginalized by social structures may seek to overturn regimes, rather than make changes within them (Ruggie 1983; Krasner 1985). Weak states that look down the decision tree and anticipate being exploited at the discretion of powerful states may even take self-damaging steps to promote their principled beliefs (Barrett 2003). Whatever the particular course of action, ideas about how the world works 'put blinders on people' and '[reduce] the number of conceivable alternatives' that they choose from (Goldstein and Keohane 1993, 12). World-views and causal beliefs, in this sense, influence issue definition, expectations, perceived interests, principled beliefs and ultimately the prospects for mutually beneficial cooperation.

In the developing world, we argue that 'structuralist' ideas about the origins and persistence of global inequality form the central world-view of most developing country leaders, including how they have viewed the issue of climate change.[12] The vast majority of goals developing country leaders sought since the end of the Second World War have remained elusive, and this, we believe, has shaped developing countries' perceptions of the world as fundamentally unequal and unjust. More than 20 years ago, Krasner (1985) argued that ideas about 'dependency' affected how many LDC decision-makers viewed the world, their identity in relation to other states, their goals and how such goals could be most effectively realized. 'The [dependency perspective] embraced by developing countries', Krasner argued, '[is] not merely a rationalization. It [is] the subjective complement to the objective condition of domestic and international weakness' (90). Najam puts it this way: '[t]he self-definition of the South . . . is a definition of exclusion: these countries believe that they have been bypassed and view themselves as existing on the periphery' (2004, 226).

There are several widely held structuralist ideas related to international environmental issues, which we argue obstruct north-south efforts to protect the climate: the idea that global environmental problems are primarily attributable to patterns of northern consumption and production, the idea that a nation's ability to implement environmental reform depends upon its position in the international division of labour and the idea that the north is using environmental issues as a ruse to thwart poor countries' economic development.[13]

---

[11] Najam (1995, 2004) explores how feelings of marginalization, frustration, anger, bitterness and hopelessness have affected bargaining positions held by developing countries, and explains that retaliatory—or 'getting even'—attitudes frequently lead to zero-sum outcomes.

[12] Through the lens of a structuralist, the international system is characterized by a division of labour. There is a global stratification system that places nations on the top, in the middle or on the bottom, and only a few manage to move up. Nations can move up or down the hierarchy, but the structure largely remains unchanged (Timmerman 1996).

[13] Najam argues that '[w]hat galvanized the North-South polarization at [the 1972 Stockholm conference and the 1992 Rio Earth Summit] was the totally different perceptions that industrialized and developing nations hold about what the "real" environmental issues exactly are' (1995, 258).

For more than 30 years, developing countries have argued that profligate consumption and production in the industrialized world is the primary driver of most global environmental problems. During ozone negotiations in the late 1980s, Brazilian Ambassador Paulo Nogueira Batista stated: '[w]e are more concerned with the threat that is coming from the North—the patterns of production and consumption of the overdeveloped countries—rather than the threat that may come from the developing countries themselves' (quoted in Sell 1996, 108). In 2008, China's Minister of Foreign Affairs used a simple analogy to explain his government's position on global climate change:

> It's like there is one person who eats three slices of bread for breakfast, and there are three people, each of whom eats only one slice. Who should be on a diet?... If per capita energy consumption is viewed in the context of the fundamental principle that people are all born equal, then I don't think some people are justified in talking about the large emissions of China, as if they have the moral high ground. (*Economic Times* 2008)

Southern policy-makers have also indicated that their ability to implement meaningful environmental reform is conditioned by their position in the international division of labour. Porter and Brown note that, '[t]he tone and substance of North-South bargaining on environmental issues are influenced by the structure of the global economic system, which exerts indirect pressure on the policies of developing countries toward their natural resources and thus constrains the quest for global cooperation to save those resources' (1991, 124; see also Chasek et al 2006).

This can be seen in both the terminology and the arguments made by developing countries. Although wealthy, industrialized countries may dismiss claims of 'environmental imperialism', 'ecological debt', 'ecologically unequal exchange' and 'environmental load displacement' as empty and distracting rhetoric, these world-views and causal beliefs shape the way southern governments view their interests.[14] Finally, a widely held perception that has endured for more than three decades is the idea that the north is willing to preserve the global environment at the expense of economic development in the south.[15]

Given these perceptions, we argue that it is not surprising southern countries have continued to bargain hard for the same opportunities afforded to now-wealthy countries at their early stages of development. As we describe in greater detail below, the 'structuralist' way of making sense of the world has promoted generalized mistrust among rich and poor nations, which in turn has suppressed diffuse reciprocity, and led to divergent and unstable expectations about future

[14] For example, in 2008, Chinese Minister of Foreign Affairs, Yang Jiechi, noted that many of China's carbon emissions are the by-product of Northern demand for manufactured goods, stating 'I hope when people use high-quality yet inexpensive Chinese products, they will also remember that China is under increasing pressure of transfer emission[s]' (*Economic Times* 2008).

[15] Najam notes that 'the South ... [sees] poverty, underdevelopment, and unequal global economic relations as the principal causes of its environmental problems and it [mistrusts] the North's environmental agenda as a guise to perpetuate this plight. It was this perception of causality of its environmental problems that informed the South's interests and strategy at both [the 1972 Stockholm and 1992 Rio] conferences' (1995, 254).

behaviour. Structuralist ideas have also promoted particularistic notions of fairness, a victim mentality and, in some cases, zero-sum or negative-sum behaviour.

Principled beliefs

The second way in which we argue global inequality influences the prospects for north-south cooperation is through its impact on the 'principled beliefs' of developed and developing countries. Goldstein and Keohane define principled beliefs as 'normative ideas that specify criteria for distinguishing right from wrong and just from unjust' (1993, 9). We argue that such ideas can facilitate cooperation if they are widely shared by providing a so-called 'focal point' that prevents endless haggling. They can also help cement cooperative solutions by increasing the willingness of 'domestic audiences' to support agreements negotiated internationally, weaken countries' incentives to hide information and misrepresent behaviour, and reduce the risk-aversion and self-damaging emotional behaviour of would-be cooperators.

Definitions of fairness are highly elastic and countries often manipulate definitions for their own narrow advantage. However, we argue that national understandings of fairness can influence the prospects for mutually beneficial cooperation by reducing the costs of negotiating, monitoring and enforcing agreements. First, norms and principles of fairness embodied in institutions can reduce the costs associated with negotiating international agreements. By providing information and lowering transaction costs, mutually acceptable 'rules of the game' reduce uncertainty, stabilize expectations, constrain opportunism, increase the credibility of state commitments and promote collective action (Keohane 1984; Abbott and Snidal 2000). For example, in so-called 'coordination dilemmas', where multiple equilibria exist along the Pareto frontier, establishing shared principles, norms, rules and decision-making procedures can enable states to zero in on a limited range of possible equilibria and enhance their prospects for cooperation (Krasner 1991). Shared understandings of fairness provide what game theorists call 'focal points'. By isolating one point along the contract curve that every party would prefer over a noncooperative outcome, states can stabilize expectations for future behaviour and reduce the costs of arriving at a mutually acceptable agreement (Keohane 2001; Müller 1999). Some theorists argue that these focal points will emerge spontaneously; others argue that when there are huge differences in fairness perceptions, states need to be explicit in identifying potential focal points for cooperation to occur (Schelling 1960; Barrett 2003; Müller 1999). That is to say, a mutually acceptable focal point will only emerge when states are willing to reconsider and negotiate their own beliefs about what is fair.[16]

The Montreal Protocol is a good example of an agreement that was guided by a fairness focal point. During the early negotiations, developed and developing countries staked out very different policy positions regarding what would constitute a 'fair' approach to combating ozone depletion (Sell 1996; DeSombre and Kauffman 1996). However, all parties eventually agreed to allow the principle

---

[16] As Müller puts it, 'we merely need a solution which is commonly regarded as sufficiently fair to remain acceptable' (1999, 3).

of 'compensatory justice' to guide the negotiations (Albin 2001; Barrett 2003). Importantly, this focal point did not spontaneously emerge from a socially shared understanding of 'appropriate' principles among countries (as the logic of social constructivism would suggest). Rather, developed and developing countries recognized that a 'negotiated justice settlement' would be necessary to hammer out a mutually acceptable agreement.[17]

In so-called 'collaboration games', where states have mixed motives for cooperation and face powerful free-rider incentives, fairness principles can also influence the costs of monitoring and enforcing agreements. Every state faces a strong temptation to free-ride on others' climate stabilization efforts in collaboration games because asymmetric information reduces the 'observability' of deviant behaviour, and the benefits of a stable climate are nonexcludable and nonrival. Thus, it is in every state's self-interest to disguise their preferences and misrepresent their level of contribution to the collective good. For these reasons, demandeurs must make compliance economically rational for more reluctant participants through financial compensation schemes, issue linkage and other forms of incentive restructuring (Young 1994, 134). This can significantly weaken incentives for cheating and defection (Krasner 1985; Abbott and Snidal 2000). Indeed, Raúl Estrada-Oyuela, one of the leading climate negotiators at Kyoto, noted that 'equity is the fundamental condition to ensure compliance of any international agreement' (Estrada-Oyuela 2002, 37).

Finally, norms and principles of fairness can help cement a collaborative equilibrium and reduce monitoring and enforcement costs through their impact on the domestic ratification process. Müller lays much emphasis on this point:

> [A skeptic] might... concede that equity has a role to play in the selection of initial allocation proposals. But surely, he is bound to interject, the *outcome* of the negotiations will be determined by good old-fashioned strategic bargaining, reflecting only the bargaining powers of the parties and the bargaining skills of the negotiators. ... [What the skeptic has overlooked is that] an agreement has to be implemented. This, in turn, requires political ratification which normally is beyond the power of mere negotiating agents.
>
> ....
>
> It would be foolish to assume, however, that bodies such as the US Congress or the Indian Lok Sabha could be ... bullied into ratifying an agreement ... [because] parties may refuse to ratify an agreement if they feel it deviates unacceptably from what they perceive to be the just solution. (1999, 10–12)

A climate of mistrust

Finally, we argue that mistrust has proven to be a major obstacle to north-south cooperation, and that this obstacle is attributable to long-standing patterns of inequality and opportunism. Inequality makes it harder for developing countries and developed countries to trust each other and establish mutually acceptable

[17] Developed and developing countries were ultimately able to find a 'middle way'—recognition that ozone depletion was a global problem requiring a global response, but that western nations had some level of responsibility to help late-developers comply with their emission reduction commitments (Sell 1996).

'rules of the game'. Such rules are important to would-be cooperators because they reduce uncertainty, stabilize expectations, constrain opportunism and increase the credibility of state commitments.

Although efforts to explore the causal impact of social trust in international environmental politics are virtually nonexistent, there is a large literature in economics, sociology and political science on the relationship between trust and cooperation (see Putnam 1993; Keohane 1984, 2001; Stein 1990; Kydd 2000). Beginning with the work of Émile Durkheim, researchers have demonstrated that by fostering norms of reciprocity, trust increases communication and information, reduces uncertainty and transaction costs, enhances the credibility of commitments, makes defection more costly, creates stable expectations and ultimately promotes cooperation (Durkheim 1933 [1893]; Putnam 1993). Trust, in effect, allows would-be cooperators to bank on promises to honour policy commitments. Putnam (1993), Knack and Keefer (1997), and Easterly (2001) and others have shown that social inequality is strongly associated with lower levels of trust, lower levels of public good provision (which is a proxy for cooperation), and higher levels of crime and other types of socially destructive behaviour.

However, there is some disagreement about how individuals, societies and nation states can best promote conditions of mutual trust. In many cases, reciprocity seems to be effective. Cicero famously said that '[t]here is no duty more indispensable than that of returning a kindness. All men distrust one forgetful of a benefit' (quoted in Putnam 1993, 172). However, as societies become more complex and information asymmetries increase, opportunities to shirk, cheat or otherwise commit malfeasance also increase, and conditions of reciprocity tend to break down. The solution to this problem in a domestic setting is simple. The creation of a state as an overarching coercive authority eliminates many perverse incentives 'by enabl[ing] its subjects to do what they cannot do on their own—trust each other' (Putnam 1993, 165). With its 'monopoly of violence', the state enforces contracts and 'coerces trust' on behalf of its citizens. However, in international relations, where contracting takes place under conditions of anarchy, states do not have the luxury of third-party enforcement (Waltz 1979; Keohane 1984; Oye 1986). They must 'decide whom to make agreements with, and on what terms, largely on the basis of their expectations about their partners' willingness and ability to keep their commitments' (Keohane 1984, 105). In the absence of a world government, rational states seeking stable and effective solutions to problems of international public good provision must develop so-called 'self-enforcing' agreements (Keohane 1984; Carraro and Siniscalco 1993; Barrett 2003).

As a result, a central question of international relations is how governments can convince potential partners that they will honour their commitments.[18] We highlight three ways in which states may seek to enhance relations of trust among themselves: specific reciprocity, diffuse reciprocity and costly signals. Specific reciprocity is when countries 'exchange items of equivalent value in a strictly delimited sequence' (Keohane 1986, 4). For example, OPEC and non-OPEC

---

[18] Mearsheimer and other 'structural realists' see 'little room for trust among states' because verifying others' promises is nearly impossible and conditions of anarchy breed constant fear (Mearsheimer 1994/1995, 11). Stein challenges this view. He argues that governments can achieve deep cooperation if they 'specify strict patterns of behavior and ensure no one cheats' (1990, 39).

nations periodically agree to cut oil production at the same time in order to maximize their impact on oil prices. However, this type of strategy has significant disadvantages. Unequal partners often find it difficult to reciprocate equally, contingencies may unexpectedly affect an actors' ability to reciprocate, and different interpretations and measurements can degenerate into situations of mutual recrimination. An accumulated stock of 'diffuse reciprocity' is much more valuable. Under circumstances where 'the definition of [equivalent value] is less precise, one's partners may be viewed as a group rather than as particular actors, and the sequence of events as less narrowly bounded' (Keohane 1986, 4); diffuse reciprocity does not require that all aspects of a contract be specified *ex ante*. Rather, it requires that states make deposits at the 'favour bank' when they can in order to build conditions of trust and stabilize expectations for future cooperative efforts (Putnam 2000, 20).

The need for diffuse reciprocity has been (indirectly) emphasized by climate policy-makers. Atiq Rahman has argued that '[t]he only way developing countries would feel comfortable getting into the system is through trust. ... Once this happens I think the ice will break' (quoted in Knight 1998). Former advisor to the Indonesian delegation Agus Sari also notes that the US decision to withdraw from Kyoto makes 'rebuilding the climate of trust ... imperative' (Sari 2003).[19] Nevertheless, when interstate relationships are characterized by mutual suspicion and deep distrust, conditions of diffuse reciprocity can be very difficult to build. As a result, states actively seeking the trust of others often need to send 'costly signals' of reassurance to disinterested or resistant parties. Such signals 'serve to separate the trustworthy types from the untrustworthy types; trustworthy types will send them, untrustworthy types will find them too risky to send' (Kydd 2000, 326). This has special relevance to the field of international environmental politics. Western countries have a long history of cooperating across a wide range of policy areas and arriving at new self-enforcing contracts, but no such history exists between developed and developing countries. North-south environmental relations are characterized by high levels of mistrust and significant power asymmetries. Najam points out that:

> [A]s a self-professed collective of the weak, the G-77 is inherently risk-averse and seeks to minimize its losses rather than to maximize its gains; ... [I]ts unity is based on a sense of shared vulnerability and a shared distrust of the prevailing world order ... [and] because of its self-perception of weakness [it] has very low expectations. (Najam 2004, 128)

Under these sorts of circumstances, the social trust literature suggests that demandeurs may need to strategically alter the payoff structures of their would-be cooperators such that voluntary participation becomes economically rational. They may also need to send special signals of reassurance that are discernible,

---

[19] Several years ago, Kevin Baumert and Nancy Kete brought scholars from the developed and developing world together through the Climate of Trust project to explore different climate policy options. They argue that '[d]espite North-South differences in emissions, wealth, and priorities, these disparities are not the largest barrier to cooperation. Probably a more potent obstacle is the enduring and growing lack of trust. Some industrialized countries have legitimate concerns that developing countries may never come into a climate protection regime or may commit only to limit their emissions at some remote future date' (2002, 10).

irrevocable, noncontingent and costly (Kydd 2000). For example, rich nations may devise compensation schemes or promote issue linkage to account for the competing policy priorities and high discount rates of developing countries (Sebenius 1983; Carraro and Siniscalco 1993).[20]

Conclusion

We now return to the north-south impasse with which we began. The Kyoto Protocol requires emission reduction commitments from a group of countries that account for approximately 19 per cent of global carbon emissions. These countries are required to reduce their emissions by a little more than 5 per cent by 2012. It is therefore widely agreed that the current global climate regime will have almost no impact on atmospheric stability. At the same time, developing country emissions are expected to rise by 60 per cent by 2030. That is not to say that Kyoto has not been useful from a political perspective, but there is a growing consensus that the focus of the international community needs to shift towards the 'post-2012' era and the central task of encouraging active southern participation in a global climate regime.

Some climate policy analysts have argued that negotiators should soldier on and continue with the same kind of horse-trading tactics that have characterized the first 15 years of climate negotiations. Widening the scope of climate negotiations to include issues like inequality, trust and fairness, in this view, will be the millstone around the neck of any effort to stabilize the climate. Instead, it is thought that western governments should invest in clean technologies, delink economic growth for carbon emissions, strengthen compliance mechanisms and proceed with or without the participation of a majority of developing countries. One might call this the 'pragmatic justice' approach—the approach that says a perfectly fair agreement existing only in the minds of negotiators is in fact unfair to all parties. However, the 'pragmatic justice' approach overlooks the daunting scientific reality that the world is already on track to seriously destabilize the atmosphere by 2020 and the south's extreme risk aversion to binding emission limits.

We argue that breaking the north-south impasse on global climate policy will likely require unconventional—and perhaps even heterodox—policy interventions. To date, countries have proposed different yardsticks for measuring atmospheric clean-up responsibilities based on particularistic notions of justice. But high levels of inequality make it very unlikely that a north-south consensus will spontaneously emerge on the basis of a single fairness principle. Therefore, a moral compromise, or 'negotiated justice' settlement, will almost certainly be necessary. To break through the cycle of mistrust that plagues north-south relations, we also argue that the north needs to offer the south a new global bargain on environment and development and signal its commitment to this new 'shared thinking' through a series of confidence building measures. Drawing upon Andrew Kydd's research on US-Soviet relations in the run-up to the end of the Cold War, we argue that a series of costly signals—'signals designed to

---

[20] During global ozone negotiations, developing countries refused to cooperate with rich nations until the threat to trade sanctions was offset by positive incentives (Barrett 2003; Mitchell and Keilbach 2001).

persuade the other side that one is trustworthy by virtue of the fact that they are so costly that one would hesitate to send them if one were untrustworthy' (Kydd 2000, 326)—can foster mutual trust between countries that do not have a long history of cooperation. These measures should offer a new vision of global environmental cooperation, provide opportunities for developing countries to transition towards less carbon-intensive development pathways and clearly signal a desire to reverse long-standing patterns of global inequality. Finally, we emphasize the central importance of exercising self-restraint when the short-term payoff on opportunistic behaviour is high. When powerful states consistently treat weaker states like second-class citizens, they run the risk of weaker states 'reciprocating' in policy domains where they possess greater bargaining leverage.

Moving towards 'hybrid justice'

Earlier, we described four very different approaches to measuring national responsibility for greenhouse gas emissions: the grandfathering approach, which relies on entitlement principles of justice; the carbon intensity approach, which rests on utilitarian principles of justice; the historical responsibility approach, which operationalizes the 'polluter pays' principle; and the egalitarian per capita approach. These particularistic notions of justice are closely associated with where countries sit in the global hierarchy of economic and political power. It is therefore very unlikely that a north-south fairness consensus will spontaneously emerge on the basis of one of these principles. Instead, a moral compromise, or 'negotiated justice' settlement, will be necessary.[21]

In recent years, a number of proposals representing moral compromise have emerged. Bartsch and Müller (2000) propose a 'preference score' method, which combines the grandfathering and per capita approach through a voting system. Their proposal allows each country—weighted by their population—to choose the methodology that they prefer. Each global citizen's 'vote' is then used to calculate national carbon emission allowances. According to their preliminary model, under this proposal, roughly three-quarters of the global emissions budget would be based on the per capita approach and one-quarter on grandfathering. Others have focused on more politically feasible per capita proposals that provide for 'national circumstances', or allowance factors, like geography, climate, energy supply and domestic economic structure, as well as 'soft landing scenarios' (see Baumert and Kete 2002; Gupta and Bhandari 1999; Agarwal et al 1999; Ybema et al 2000; Ringus et al 2002; Torvanger and Ringius 2002; Torvanger and Godal 2004).

The Pew Center for Global Climate Change has developed a hybrid proposal that assigns responsibility based on past and present emissions, carbon intensity and countries' ability to pay (that is its per capita GDP) (Claussen and McNeilly 1998). It separates the world into three groups: those that 'must act now', those that 'could act now' and those that 'should act now, but differently'. The 'Triptych' proposal, designed by scholars at the University of Utrecht (and already used to

[21] This point is increasingly recognized by scholars and policy-makers. Blanchard et al note that 'any future burden-sharing agreement involving developing countries will probably be based on a complex differentiation scheme combining different basic rules' (2003, 286).

differentiate commitments among EU countries), 'accounts for differences in national circumstances such as population size and growth, standard of living, economic structure and fuel mix in power generation' (Groenenberg et al 2001). Its novel contribution is that it divides each country's economy into three sectors: energy-intensive industry, power generation and the so-called domestic sector (transport, light industry, agriculture, and commercial sector). It applies the carbon intensity approach to the energy-intensive sector, 'decarbonization targets' to the power generation sector and a per capita approach to the 'domestic' sectors. Similarly, the multisector convergence approach, developed by two research institutes in northern Europe (ECN and CICERO), treats sectors differentially and integrates per capita, carbon intensity and ability to pay (GDP per capita) approaches (Sijm et al 2000).

EcoEquity.org has also created a 'Greenhouse Development Rights' framework as a reference to evaluate proposals for the post-2012 period (Baer et al 2007). They argue that countries below a 'global middle class' income of 9,000 USD per capita should be assured that they will not be asked to make binding limits until they approach that level; countries above that level should be responsible for rapid reductions of emissions and payments to assist those below the line in improving their social and economic status while adjusting to a less carbon-intensive path of development. We believe these hybrid proposals are among the most promising solutions to break the north-south stalemate.

Building trust through costly signals

We also argued that international climate negotiations are hobbled by the wider 'trust deficit' that plagues north-south environmental relations, and that a strategy of reassurance through costly signals can foster mutual trust between countries that do not have a long history of cooperation. In particular, we believe Kydd's (2000) research on US-Soviet relations during the 1980s offers important lessons for students of north-south environmental politics. He argues that the more noticeable, irreversible, unconditional and costly the signal from the sending state, the more trusting the receiving state will be and willing to engage in cooperative ventures.

> [I]n looking at the end of the Cold War ... [we] can observe a series of costly signals leading to mutual trust between former adversaries. The attitudes of Western leaders, press, and publics toward the Soviet Union all underwent a substantial transformation. Soviet military and geopolitical concessions, particularly the [Intermediate-range Nuclear Forces] treaty, the withdrawal from Afghanistan, the December 1988 conventional arms initiative, and the withdrawal from Eastern Europe were decisive in changing overall Western opinion about the Soviet Union. By 1990 most observers viewed the Soviet Union as a state that had abandoned its hegemonic ambitions and could be trusted to abide by reasonably verified arms control agreements and play a constructive role in world politics. (350)

Importantly, this process of trust-building is usually an incremental and long-term strategy that involves sending a series of costly signals—and, in some cases, progressively more costly signals.

In a sense, the conditions of mistrust which currently plague north-south environmental relations can be understood as the product of a 'failed reassurance strategy'. In the early 1990s, the north assured poorer nations that they would 'take the lead' in stabilizing the climate (UNFCCC 1992, article 3.1). However, subsequent efforts have been sluggish, litigious, uneven and generally unimpressive. The lack of progress by the US and other industrialized nations in meeting their own emission reduction targets has provided developing nations a ready excuse for not making cuts. As Brazil's leading newspaper put it, 'Numbers like these [the US's emissions] reinforce the disposition of the Brazilian government to reject the idea of taking on additional costs to do its part in reducing the greenhouse effect' (Rossi 1997).

However, it is important to note that some trust-building efforts have proven successful in global environmental politics. The Multilateral Ozone Fund and the reformed Global Environmental Facility enshrined the 'compensatory justice' principle and gave developing countries a greater stake in the decision-making process governing the allocation of environmental aid (Woods 1999). The Montreal Protocol also gave developing countries a ten-year window to pursue 'cheap' economic development before making serious chlorofluorocarbon reductions. In the climate change arena, the way has begun to be paved with the three special funds set up at COP6 in Bonn and COP7 in Marrakesh to help developing countries adapt to the adverse effects of climate change, facilitate technology transfer and mitigate greenhouse gas emissions. Developed countries have also served themselves well by inviting developing countries to participate in the 'Compliance Committee' (without having to adopt scheduled emission reduction commitments) and treating them as equal partners through the double-majority voting mechanism.[22]

Strategic restraint: the second prong of a trust-building strategy

Trust-building is of course not only about what you do, but also exercising strategic restraint. As we have argued elsewhere, one important way to demonstrate solidarity, empathy, kindness, friendship and loyalty to developing countries is to acknowledge and support their concerns and priorities in the international political economy arena. In fact, this could ultimately prove to be *more* important than international treaties, carbon accounting schemes or environmental aid. According to seasoned analyst Herman Ott, 'it became clear [at COP8 in New Delhi] that developing countries would not give up their "right" for increasing emissions without serious concessions in other fields of the development agenda which satisfy the demand for global equity and poverty reduction' (2004, 261).

Scholars of environmental politics unfamiliar with the international political economy literature may view such demands as distracting and unconstructive, but the ongoing development crisis is at the very heart of the climate policy gridlock. Developing countries want more 'policy space'—room to define and pursue their own development agenda—but contemporary international economic regimes present huge hurdles to export diversification, institutional experimentation and upward mobility in the world economy (Wade 2003).

[22] A double majority voting system requires that two separate majorities be achieved—in this case, a majority of the wealthy, industrialized 'Annex I' countries and a majority of 'Non-Annex I' countries.

As such, if industrialized nations are interested in securing a north-south global climate agreement, we argue that they should consider explicitly signalling their concern for the 'structural obstacles' facing developing countries. For example, the current practice of tariff escalation reinforces the structuralist perception that rich countries do not want poor countries to get rich the same way they did. The TRIPS agreement has a similar effect since rich nations historically had complete policy autonomy in this area, granting patents at their own discretion in order to encourage industrial transformation (Shadlen 2004). Other possibilities include reining in the 'deep integration' and anti-industrial policy crusade, not punishing poor countries for export diversification efforts, recognizing that the 'political losers' created by the diversification process must be somehow compensated and promoting predictability (and reducing opportunities for opportunism) in international economic regimes (Wade 2003; Birdsall et al 2005; Shadlen 2004).

Working towards a 'new shared thinking' about north-south environmental relations

Finally, we documented the existence of an enormous 'world-view gap' between rich and poor nations and argued that the 'structuralist' world-views and causal beliefs of developing countries will persist unless western governments make efforts to reverse this inequality. Yet, under circumstances of extreme mistrust, Kydd (2000) suggests that reluctant states may require more than costly signals and strategic reassurance. It may also be necessary for would-be cooperators to establish a 'shared world-view'. Kydd argues that this is typically the result of one state (or group of states) trying to get another state (or group of states) to buy into a so-called 'new thinking'. For example, during the Cold War, the US and Soviet administrations worked together to establish a 'new thinking' about global security. When the Soviet Union withdrew from Afghanistan, an editorial appeared in *The New York Times*, noting that its actions 'begin to render credible Moscow's "new thinking" about the Soviet role in the world' (quoted in Kydd 2000, 346). Subsequently, when 'asked if he still held to the idea that the Soviet Union was an evil empire ... [Reagan] responded, "No, I was talking about another time, another era."' (346).

Athanasiou and Baer suggest that, in the context of international climate negotiations, the greatest challenge is to ensure that the south 'not [view] climate justice as the justice of following the North down the fossil-fuel path' (2002, 83). Wheeler and Ummel similarly note that policy-makers need to be disabused of 'the notion that the South can utilize carbon-intensive growth to dramatically increase incomes—a kind of last-minute, fossil-fuelled development push—before the onset of catastrophic climate change' (2007, 9). But to move away from 'old North-South thinking', or what Graham calls 'residual 1970s thinking' (1996, 216), an attractive alternative must be offered. We argue that the north will need to aggressively assist developing countries in making the tough transition to more lower carbon pathways of development.[23] This is not merely a financing issue.

[23] Wheeler and Ummel echo this point: 'The North must ... [recognize] that its own survival requires an immediate, large-scale commitment to assisting emissions reductions in the South' (2007, 11). Also see Hicks et al (2008).

Countries on high-emission pathways will require serious attention to their political and class situations, as diversification is an intensely political process and conflict will inevitably arise. Developing countries will therefore need 'policy space' to pursue strategies tailored to local culture, knowledge, institutions and politics, while being provided significant technical assistance, technology transfer and aid.

To conclude, climate change is fundamentally an issue of inequality and its resolution will necessitate an unconventional policy approach. Climate negotiations, we must remember, take place in the context of an ongoing development crisis and what is perceived by the global south as a pattern of northern callousness and opportunism in matters of international political economy. They take place at a time when levels of generalized trust are declining. And they take place at a time when poor nations' concerns for fair processes and outcomes have frequently been marginalized. Negotiators must therefore explicitly and aggressively signal concern and seek to address the structural obstacles facing developing countries. We need a global and just transition built on diffuse reciprocity, a climate of trust, negotiated justice and a shared world-view.

References

Abbott, Kenneth and Duncan Snidal (2000) 'Hard and soft law in international governance', *International Organization*, 54:3, 421–456

Agarwal, Anil, Sunita Narain, Anju Sharma and Achila Imchen (2001) *Poles apart: global environmental negotiation-2* (New Delhi, India: Centre for Science and Environment)

Albin, Cecilia (2001) *Justice and fairness in international negotiation* (Cambridge, United Kingdom: Cambridge University Press)

Athanasiou, Tom and Paul Baer (2002) *Dead heat: global justice and global warming* (New York: Seven Stories Press)

Baer, Paul, Thomas Athanasiou and Sivan Kartha (2007) 'The right to development in a climate constrained world: the greenhouse development rights framework', Report of EcoEquity and the Stockholm Environment Institute, <http://www.ecoequity.org/docs/TheGDRsFramework.pdf>, accessed 10 July 2008

Barrett, Scott (2003) *Environment and statecraft: the strategy of environmental treaty-making* (Oxford, United Kingdom: Oxford University Press)

Bartsch, Ulrich and Benito Müller (2000) *Fossil fuels in a changing climate: impacts of the Kyoto Protocol and developing country participation* (Oxford, United Kingdom: Oxford University Press)

Baumert, Kevin A (ed) (2002) *Building on the Kyoto Protocol: options for protecting the climate* (Washington, DC: World Resources Institute)

Baumert, Kevin A and Nancy Kete (2002) 'An architecture for climate protection' in Kevin Baumert (ed) *Building on the Kyoto Protocol: options for protecting the climate* (Washington, DC: World Resources Institute), 1–30

Birdsall, Nancy, Dani Rodrik and Arvind Subramanian (2005) 'If rich governments really cared about development', Working Paper, International Centre for Trade and Sustainable Development, Geneva

Blanchard, Odile, Patrick Criqui, Alban Kitous and Laurent Viguier (2003) 'Combining effficiency with equity: a pragmatic approach' in Inge Kaul, Pedro Conceição, Katell Le Goulven and Ronald U Mendoza (eds) *Providing global public goods: managing globalization* (New York: Oxford University Press)

Carraro, Carlo and Domenico Siniscalco (1993) 'Strategies for the international protection of the environment', *Journal of Public Economics*, 52:3, 309–328

Chasek, Pamela, David Downie and Janet Welsh Brown (2006) *Global environmental politics*, fourth edition (Boulder, Colorado: Westview Press)

Claussen, Eileen and Lisa McNeilly (1998) *Equity and global climate change: the complex elements of fairness* (Arlington, Virginia: Pew Center on Climate Change)

Depledge, Joanna (2002) 'Continuing Kyoto: extending absolute emission caps to developing countries' in Kevin Baumert (ed) *Building on the Kyoto Protocol: options for protecting the climate* (Washington, DC: World Resources Institute), 31–60

DeSombre, Elizabeth R and Joanne Kauffman (1996) 'The Montreal Protocol Multilateral Fund: partial success story' in Robert O Keohane and Marc A Levy (eds) *Institutions for environmental aid: pitfalls and promise* (Cambridge, Massachusetts: MIT Press), 89–126

Dessai, Suraje (2001) 'Why did The Hague climate conference fail?', *Environmental Politics*, 10:3, 139–144

Durkheim, Emile (1933 [1893]) *The division of labor in society* (New York: Free Press)

Easterly, William (2001) 'The middle class consensus and economic development', *Journal of Economic Growth*, 6:4, 317–335

Economic Times (2008) 'China tells developed world to go on climate change "diet"', 12 March

Estrada-Oyuela, Rául A (2002) 'Equity and climate change' in Luiz Pinguelli-Rosa and Mohan Munasinghe (eds) *Ethics, equity and international negotiations on climate change* (Cheltenham, United Kingdom: Edward Elgar Publishing), 36–46

European Union (2008) 'Climate change and international security', Paper from the High Representative and the European Commission to the European Council, Brussels, 14 March <http://www.consilium.europa.eu/ueDocs/cms_Data/docs/pressData/en/reports/99387.pdf> , accessed 10 July 2008

Ford, Peter (2007) 'China balks at emissions caps', *Christian Science Monitor*, 5 June <http://www.csmonitor.com/2007/0605/p01s04-woap.html>

Goddard Institute for Space Studies (2008) 'Global temperature anomalies in .01 C', <http://data.giss.nasa.gov/gistemp/tabledata/GLB.Ts.txt> , accessed 10 July 2008

Goldstein, Judith and Robert Keohane (eds) (1993) *Ideas and foreign policy: beliefs, institutions, and political change* (Ithaca, New York: Cornell University Press)

Graham, Edward M (1996) 'Direct investment and the future agenda of the World Trade Organization' in Jeffrey J Schott (ed) *The world trading system: challenges ahead* (Washington, DC: Institute for International Economics), 205–217

Groenenberg, Heleen, Dian Phylipsen and Kornelis Blok (2001) 'Differentiating commitments world wide: global differentiation of GHG emissions reductions based on the Triptych approach-a preliminary assessment', *Energy Policy*, 29:12, 1007–1030

Gruber, Lloyd (2000) *Ruling the world: power politics and the rise of supranational institutions* (Princeton, New Jersey: Princeton University Press)

Gupta, Joyeeta (2000) *'On behalf of my delegation, ...': a survival guide for developing country climate negotiators* (Washington, DC: Center for Sustainable Development in the Americas)

Gupta, Sujata and Preety M Bhandari (1999) 'An effective allocation criterion for CO_2 emissions', *Energy Policy*, 27:12, 727–736

Haas, Peter, Robert Keohane and Marc Keohane (eds) (1993) *Institutions for the earth: sources of effective international environmental protection* (Cambridge, Massachusetts: MIT Press)

Haas, Peter M (1990) *Saving the Mediterranean: the politics of international environmental cooperation* (New York: Columbia University Press)

Heil, Mark T and Thomas M Selden (2001) 'International trade intensity and carbon emissions: a cross-country econometric analysis', *Journal of Environment and Development*, 10:1, 35–49

Hicks, Robert L, Bradley C Parks, J Timmons Roberts and Michael J Tierney (2008) *Greening aid? Understanding the environmental impact of development assistance* (Oxford, United Kingdom: Oxford University Press)

IPCC (2007) *Climate change 2007: fourth assessment report of the Intergovernmental Panel on Climate Change* (Cambridge, United Kingdom: Cambridge University Press)

Jeter, Jon (2002) 'Mistrust reigns as second Earth Summit commences', *Washington Post*, 26 August

Kaul, Inge, Isabelle Grunberg and Marc Stern (1999) 'Defining global public goods' in Inge Kaul, Isabelle Grunberg and Marc Stern (eds) *Global public goods: international cooperation in the 21st century* (Oxford, United Kingdom: Oxford University Press), 2–19

Keck, Margaret E and Kathryn Sikkink (1998) *Activists beyond borders: advocacy networks in international politics* (Ithaca, New York: Cornell University Press)

Keohane, Robert (1984) *After hegemony: cooperation and discord in the world political economy* (Princeton, New Jersey: Princeton University Press)

Keohane, Robert (1986) 'Reciprocity in international relations', *International Organization*, 40:1, 1–27

Keohane, Robert (2001) 'Governance in a partially globalized world', *American Political Science Review*, 95:1, 1–13

Keohane, Robert and Marc A Levy (eds) (1996) *Institutions for environmental aid: pitfalls and promise* (Cambridge, Massachusetts: MIT Press)

Knack, Stephen and Philip Keefer (1997) 'Does social capital have an economic payoff? A cross-country investigation', *Quarterly Journal of Economics*, 112:4, 1251–1288

Knight, Danielle (1998) 'Environment: in defence of developing nations', *IPS News*, 27 October, <http://www.sunsonline.org/trade/process/followup/1998/10270498.htm>

Krasner, Stephen (1978) 'United States commercial and monetary policy: unraveling the paradox of external strength and internal weakness' in Peter J Katzenstein (ed) *Between power and plenty: foreign economic policies of advanced industrial states* (Madison: University of Wisconsin Press), 51–87

Krasner, Stephen (1985) *Structural conflict: the third world against global liberalism* (Berkeley and Los Angeles, California: University of California Press)

Krasner, Stephen D (1991) 'Global communications and national power: life on the Pareto Frontier', *World Politics*, 43:3, 336–366

Kydd, Andrew (2000) 'Trust, reassurance, and cooperation', *International Organization*, 54:2, 325–357

Levy, David L and Ans Kolk (2002) 'Strategic responses to global climate change: conflicting pressures on multinationals in the oil industry', *Business and Politics*, 4:2, 275–300

Machado, Giovani, Roberto Schaeffer and Ernst Worrell (2001) 'Energy and carbon embodied in the international trade of Brazil: an input-output approach', *Ecological Economics*, 39:3, 409–424

Mathur, Ajay, Ian Burton and Maarten van Aalst (eds) (2004) *An adaptation mosaic: a sample of the emerging World Bank work in climate change adaptation* (Washington, DC: World Bank)

Mearsheimer, John J (1994/1995) 'The false promise of international institutions', *International Security*, 19:3, 5–49

Meyer, John W, David John Frank, Ann Hironaka, Evan Schofer and Nancy Brandon Tuma (1997) 'The structuring of a world environmental regime, 1870–1990', *International Organization*, 51:4, 623–629

Mitchell, Ronald B and Patricia Keilbach (2001) 'Reciprocity, coercion, or exchange: symmetry, asymmetry and power in institutional design', *International Organization*, 55:4, 893–919

Müller, Benito (1999) *Justice in global warming negotiations: how to obtain a procedurally fair compromise* (Oxford, United Kingdom: Oxford Institute for Energy Studies)

Müller, Benito (2001) 'Fair compromise in a morally complex world', Paper presented at Pew Equity Conference, Washington, DC, 17–18 April

Muradian, Roldan, Martin O'Connor and Joan Martinez-Alier (2002) 'Embodied pollution in trade: estimating the "environmental load displacement" of industrialized countries', *Ecological Economics*, 41:1, 51–67

Najam, Adil (1995) 'International environmental negotiations: a strategy for the south', *International Environmental Affairs*, 7:2, 249–287

Najam, Adil (2002) 'Financing sustainable development: crises of legitimacy', *Progress in Development Studies*, 2:2, 153–160

Najam, Adil (2003) 'The case against a new international environmental organization', *Global Governance*, 9:3, 367–384

Najam, Adil (2004) 'The view from the south: developing countries in global environmental politics' in Regina Axelrod, David Downie and Norman Vig (eds) *The global environment: institutions, law, and policy*, second edition (Washington DC: CQ Press), 225–243

Ott, Herman (2004) 'Global climate' in Geir Ulfstein and Jacob Werksman (eds) *Yearbook of international environmental law 12* (Oxford, United Kingdom: Oxford University Press), 261–270

Oye, Kenneth A (ed) (1986) *Cooperation under anarchy* (Princeton, New Jersey: Princeton University Press)

Porter, Gareth and Janet W Brown (1991) *Global environmental politics* (Boulder, Colorado: Westview Press)

Putnam, Robert D (1993) *Making democracy work* (Princeton, New Jersey: Princeton University Press)

Putnam, Robert (2000) *Bowling alone: The collapse and revival of American community* (New York: Simon and Schuster)

Raustiala, Kal (1997) 'Domestic institutions and regulatory cooperation: comparative responses to the Global Biodiversity Regime', *World Politics*, 49:4, 482–509

Ringus, Lasse, Asbjorn Torvanger and Arild Underdal (2002) 'Burden sharing and fairness principles in international climate policy', *International Environmental Agreements: Politics, Law and Economics*, 2:1, 1–22

Roberts, J Timmons and Bradley C Parks (2007) *A climate of injustice: global inequality, north-south politics, and climate policy* (Cambridge, Massachusetts: MIT Press)

Robinson, Nicholas A (ed) (1992) *Agenda 21 and UNCED proceedings, volumes 1 and 2* (New York: Oceana Publications)

Rodrik, Dani (2001) 'Trading in illusions', *Foreign Policy*, 123 (March-April), 54–62

Rossi, Clóvis (1997) 'Controle da emissno de gases divide FHC e Clinton [Control of emission gases divides Clinton and the FHC]', *Folha de Sao Paulo*, 10 October, A-15

Ruggie, John Gerard (1983) 'Political structure and change in the international economic order: the north-south dimension' in John Gerard Ruggie (ed) *The antinomies of interdependence* (New York: Columbia University Press), 423–487

Sandbrook, Richard (1997) 'UNGASS has run out of steam', *International Affairs*, 73:4, 641–654

Sari, Agus (2003) 'CoP-9 in Milan: a fashionable trend for climate agreements', Centre for Science and Environment, <http://www.cseindia.org/campaign/ew/agreements.htm>, accessed 10 July 2008

Schelling, Thomas (1960) *The strategy of conflict* (Cambridge, Massachusetts: Harvard University Press)

Sebenius, JK (1983) 'Negotiation arithmetic: adding and subtracting issues and parties', *International Organization*, 37:2, 281–316

Sebenius, James K (1991) 'Designing negotiations towards a new regime: the case of global warming', *International Security*, 15:4, 110–148

Sell, Susan (1996) 'North-south environmental bargaining: ozone, climate change, and biodiversity', *Global Governance*, 2:1, 97–118

Shadlen, Ken (2004) 'Patents and pills, power and procedure: the north-south politics of public health in the WTO', *Studies in Comparative International Development*, 39:3, 76–108

Sijm, Jos PM, Jaap C Jansen, JJ Battjes, CH Volkers and JR Ybema (2000) *The multi-sector convergence approach of burden sharing: an analysis of its cost implications* (Oslo, Norway: Center for International Climate and Environmental Research)

Sprinz, Detlef and Tapani Vaahtoranta (1994) 'The interest-based explanation of international environmental policy', *International Organization*, 48:1, 77–105

Stavins, Robert and Sheila M Olmstead (2006) 'An international policy architecture for the post-Kyoto era', *American Economic Review Papers and Proceedings*, 96:2, 35–38

Stein, Arthur (1990) *Why nations cooperate: circumstance and choice in international relations* (Ithaca, New York: Cornell University Press)

Timmerman, P (1996) 'Breathing room: negotiations on climate change' in FO Hampson and J Reppy (eds) *Earthly goods: environmental change and social justice* (Ithaca, New York: Cornell University Press), 221–243

Torvanger, Asbjorn and Odd Godal (2004) 'An evaluation of pre-Kyoto differentiation proposals for national greenhouse gas abatement targets', *International Environmental Agreements: Politics, Law and Economics*, 4:1, 65–91

Torvanger, Asbjorn and Lasse Ringius (2002) 'Criteria for evaluation of burden-sharing rules in international climate policy', *International Environmental Agreements: Politics, Law and Economics*, 2:3, 221–235

UNFCCC (1992) *United Nations Framework Convention on Climate Change* (New York: United Nations), <http://unfccc.int/resource/docs/convkp/conveng.pdf>, accessed 10 July 2008

Victor, David (2001) *The collapse of the Kyoto Protocol and the struggle to slow global warming* (Princeton, New Jersey: Princeton University Press)

Wade, Robert (2003) 'What strategies are viable for developing countries today? The World Trade Organization and the shrinking of development space', *Review of International Political Economy*, 10:4, 627–644

Waltz, Kenneth (1979) *Theory of international politics* (New York: Random House)

Wapner, Paul (1995) 'Politics beyond the state: environmental activism and world civic politics', *World Politics*, 47:3, 311–340

Wheeler, David and Kevin Ummel (2007) 'Another inconvenient truth: a carbon-intensive south faces environmental disaster no matter what the north does', Working Paper 134, Center for Global Development, Washington, DC

Woods, Ngaire (1999) 'Good governance in international organizations', *Global Governance*, 5:1, 36–61

Ybema, J Remko, JJ Battjes, Jaap C Jansen and Frank Ormel (2000) *Burden differentiation: GHG emissions, undercurrents and mitigation costs* (Oslo, Norway: Center for International Climate and Environmental Research)

Young, Oran R (1994) *International governance: protecting the environment in a stateless society* (Ithaca, New York: Cornell University Press)

Greenhouse Development Rights: towards an equitable framework for global climate policy

Paul Baer, Glenn Fieldman, Tom Athanasiou and Sivan Kartha[1]

Introduction

Since the 1992 adoption of the United Nations Framework Convention on Climate Change (UNFCCC), whose core mandate is to 'prevent dangerous anthropogenic interference with the climate system', the concentration of atmospheric carbon dioxide (CO_2) has risen dramatically and the *rate* at which the concentration is rising has accelerated in recent years (Canadell et al 2007). During the same time

[1] The authors gratefully acknowledge the support of many colleagues who have contributed to the development of the ideas in this article. Eric Kemp-Benedict of the Stockholm Environment Institute has provided invaluable support both to our calculations and our analysis of inequality within countries. Financial support for Greenhouse Development Rights has come from Mistra's Climate Policy Research Program, Christian Aid, the Heinrich Böll Foundation, the Town Creek Foundation, Oxfam International, and the Stockholm Environment Institute. We also thank several anonymous reviewers whose comments greatly helped this article. All remaining errors of fact and opinion are strictly our own. Periodic updates to the GDRs calculations can be found at www.ecoequity.org/GDRs

period evidence has mounted that the measured increase in global average temperature over the last century—caused substantially by CO_2 and other greenhouse gases (GHGs) added to the atmosphere since the Industrial Revolution—is resulting in adverse effects to ecosystems, precipitating glacier melt in most mountainous regions and beginning to disintegrate the ice cover at both poles.[2] Scientific evidence strongly supports the case not only for GHG emissions cuts, but for *deep* cuts in a matter of decades;[3] the interests to be protected here are not just those of future generations, but those of people now living.

The Kyoto Protocol to the Convention, which came into effect in 2005, required obligated countries to make only modest emissions cuts. It has failed even to significantly reduce the rate at which global emissions are rising—nor is it expected to do so even if all ratifying parties meet their obligations under the Protocol by the end of the commitment period in 2012. This is because the Protocol, though an important first step in addressing the UNFCCC mandate because it includes binding reduction commitments, lacks some essential features. First, although it calls for reductions on the part of 'Annex I' (industrialized) signatories below an emission baseline set in 1990, the Protocol does not include scientifically informed specific temperature targets or corresponding atmospheric carbon concentration targets: in other words, it does not include any criteria for what might be called 'climate adequacy', by which we mean emissions cuts that are calibrated to some defined threshold of 'dangerous climate change'.

Second, the Protocol was crafted in accord with the UNFCCC division of the world into Annex I (obligated industrial countries) and non-Annex I (developing, nonobligated countries). This division operationalizes the UNFCCC requirement that commitments be reflective of unequal national responsibilities and capabilities, and represents a crucial recognition of global inequality. But the Protocol's criteria have been criticized as arbitrary, and the binary division inadequate to take account of what is actually a vast range of difference in national circumstances. Moreover, because non-Annex I countries are exempt from mitigation commitments, the Protocol does not directly address the large and growing aggregate emissions (albeit still small in comparative per capita terms) of rapidly industrializing countries such as China and India. The exemption of these large aggregate emitters under Kyoto has contributed substantially to the political deadlock evident at Kyoto and in subsequent negotiations, since it has been used by the United States (US), the largest aggregate emitter up to the present, to justify its refusal to ratify Kyoto or undertake *any* binding international mitigation commitment.

Our primary purpose in this article is to present Greenhouse Development Rights (GDRs) as a principled equitable framework by which a precautionary

[2] The scientific literature on impacts through roughly 2005 was summarized by the Intergovernmental Panel on Climate Change (IPCC) in its Fourth Assessment Report (see for example the Synthesis Report; IPCC 2007a) but worrisome new reports continue to appear regularly (see, for example, *Science Daily* 2007).

[3] Advocates of precautionary climate policy have long cited 2°C above the preindustrial mean as a threshold for 'dangerous climate change' to be avoided; even a 50 per cent chance of staying below this level requires emissions to peak by 2020 and drop to at least 50 per cent below 1990 levels by 2050 (IPCC 2007b, 15–16). Evidence of significant harm from the current temperature increase of 0.8°C now suggests that even the 2°C threshold may be unacceptably high—see for example, Hansen (2007) and Spratt and Sutton (2008).

emissions budget (total allowable emissions over a given time period) might be divided, and which suggests how the costs of mitigation and of adaptation to climate-induced changes that cannot be avoided might be fairly shared. A just distribution of emissions cuts, their costs and the costs of adaptation, we argue, would impose the greatest burden on wealthy people and countries, while preserving for poor people and poor countries the means to achieve 'sufficiency' (Page 2006, 85–93). We believe such a distribution of costs to be equitable. Without the right to the majority of the world's remaining emissions budget and substantial financial assistance for low-carbon development, the world's majority population will likely find its opportunity to enjoy a decent standard of living foreclosed because of the well-documented historical relationship between prosperity and per capita energy use.

A principled and transparent division of mitigation obligations and costs has been called for by a number of developing country parties to the UNFCCC. At the 2007 UNFCCC meeting in Bali, representatives from Africa, South Asia, and Central America underlined the need to take account of particular national circumstances (UNFCCC 2008a, 15). We believe that a principled and transparent approach to 'burden sharing'—or, as some call it, 'differentiation of commitments'—offers the best chance to bring high aggregate-emissions developing countries such as China and India into a post-Kyoto climate regime. These countries must be on board if the world is to achieve a precautionary emissions target within the necessary time frame. A framework that explicitly guarantees the right to development—that is, atmospheric space, the financial and technological means to provide energy services, and time to grow without Draconian restrictions —is thus an evident political necessity.

The GDRs framework is conceived as a design that can satisfy the 'adequacy and equity imperatives' that were enshrined although not specified in the UNFCCC. Sovereign states claim to seek an adequate and equitable solution to the climate problem; the GDRs framework suggests what such a solution might look like, and provides a transparent formula and set of criteria through which the obligations of states can be quantified. Thus GDRs are not a mere thought experiment: the imperatives have been officially recognized and agreed to. The first part of our article therefore lays out precautionary criteria for adequacy, quantifies the conditions that would enable the satisfaction of the equity imperative and explains the formula from which we derive the suggested obligations of states. A unique, and indeed, central feature of the GDRs framework is that the distribution of income *within* states is used as one of the bases upon which their obligations are calculated. This is a novel and potentially controversial feature, but it reflects much more accurately the actual global income distribution (and thus capacity to pay) than any framework that divides obligations among rich and poor *countries*.

Second, we suggest ways that the framework might be relevant to the present negotiations for the second and subsequent commitment periods of the Kyoto Protocol,[4] and more generally to the coming decades of international

[4] The quantified obligations of the Kyoto Protocol govern the so-called 'First Commitment Period' from 2008 to 2012. A second commitment period would thus presumably begin in 2013 but its end date would be negotiated.

climate policy. As we will show, the GDRs framework offers a way to rectify the deficiencies in the binary (Annex I/non-Annex I) differentiation of commitments in the Kyoto Protocol, and to calculate mitigation obligations appropriate to the continuum of countries in terms of their capacity and responsibility. However, a transition to a full global cap-and-trade or equivalent scheme may not immediately be feasible regardless of the allocation principles; thus we address the potential role of GDRs in an incremental evolution of the Kyoto and UNFCCC architectures.

The Greenhouse Development Rights framework

The GDRs framework is a global climate policy architecture for the allocation of costs—a *burden-sharing* framework, in the policy jargon. It offers a transparent and defensibly fair method for allocating costs, whether for mitigation or adaptation, based on principles of *capacity* and *responsibility,* both of which are widely acknowledged ethical criteria and both of which are regularly invoked by parties to international climate negotiations.[5] Crucially, the framework quantifies national obligations on the basis of the income distribution within countries, so that conceptually a person with the same income and historical emissions would bear the same proportional share of global obligations, regardless of the country in which he or she lives. We believe this is an appropriate ethical principle for the allocation of obligations in a global community, that it embodies a notion of 'fairness' that is understandable and transparent, and is therefore politically defensible.

Although the GDRs framework is in some sense independent of the particular goal that is set for a precautionary global emissions budget, it is motivated in large part by the recognition that the urgency of the climate crisis requires global emissions to fall rapidly long before today's poor countries become 'developed'. In this context, in the absence of a true technological miracle, we see only three possibilities: the climate regime severely limits the growth in energy use of the world's poor; the rich pay to subsidize low-carbon development globally; or we abandon the possibility of rapid emissions reductions that can hope to stabilize the climate. The dilemma is illustrated by Figure 1, which shows a global 'emergency pathway' emissions trajectory that peaks in 2013 and falls at about 5.5 per cent annually to 80 per cent below 1990 levels by 2050, and which *still* has a roughly 14 to 32 per cent chance of leading to temperature increase of more than 2°C above the preindustrial mean.[6] Against this we show an 'optimistic' pathway for the current Annex I countries (bottom line), in which their emissions drop to

[5] Indeed, the UNFCCC itself states that countries should 'protect the climate system... on the basis of equity and in accordance with their common but differentiated responsibilities and respective capabilities' (UNFCCC 1992, article 3 paragraph 1).

[6] The likelihood that any emissions pathway will lead to temperature increase over, for example, 2°C is not an objective probability (like the probability of rolling a six on a fair die). Estimates of these kinds are necessarily *subjective probabilities* and, given the complexity of the system and our ignorance about its mechanisms, reasonable people will disagree. Thus these kinds of risk estimates can only be reported as 'ranges of reasonable probabilities'. These ranges here were calculated with the Monte Carlo Climate Model (MC^2), calibrated to results of GCMs published in the scientific literature; calculations based on different models would presumably overlap the ranges reported here. See Baer and Mastrandrea (2006).

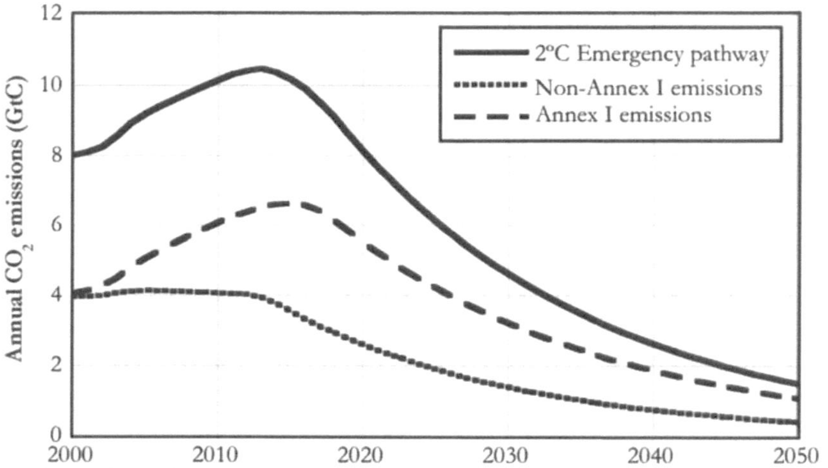

Figure 1. Hypothetical CO_2 budgets for Annex I and non-Annex I countries under global 'emergency pathway' (see main text)

90 per cent below 1990 levels by 2050. What is left by subtraction is the available carbon budget for the non-Annex I countries, which is represented by the line in the middle. What Figure 1 demonstrates is that emissions in non-Annex I countries must begin to fall shortly after the global peak (2015 in this calculation) and must soon thereafter fall at almost the same rate as global emissions.

What Figure 1 does *not* show is the development status of the non-Annex I countries at the time their emissions must peak and drop. Although such projections are obviously speculative, the fact that the non-Annex I peak must occur within a decade clarifies the situation: even at 5 per cent annual growth in per capita incomes,[7] the average income in the non-Annex I countries in 2018 would still be less than $7,500, adjusted by Purchasing Power Parity (PPP).[8] Furthermore, on projected growth rates, per capita *emissions* in the developing countries would be only about 1.1tC (ton of carbon) per annum, compared to current levels in the Annex I countries of over 3tC. In this context, the idea that developing countries would agree to pay to mitigate at the same rate as developed countries seems, to say the least, far fetched.

Yet, these *average* projected incomes and emissions in the developing countries mask enormous differences of income between them—as well as unequal income distributions *within* them (we leave aside for now the parallel issue in developed countries). At least ten non-Annex I countries have per capita incomes over

[7] For comparison, the countries classified as 'low income' by the World Bank, with roughly 35 per cent of global population, have seen real per capita income growth (in PPP terms) averaging 4.6 per cent annually over the last five years and 3.2 per cent annually over the last 20 years; the lower-middle income group, with roughly the same population (and including China) has had growth of 5.8 per cent annually over the last five years and 5.0 per cent over the last 20 years.

[8] This translates the purchasing power of various national currencies into purchasing power in US dollars. Because a PPP-adjusted figure shows what local incomes *actually buy* in dollar terms, PPP adjustments facilitate more accurate comparisons between countries than conversion using market exchange rates.

$20,000 annually—it is a stretch to still call them 'developing'.[9] And across all of non-Annex I, our estimate suggests that there are already about 750 million people whose incomes exceed $7,500 per capita (PPP adjusted), a level roughly equal to the poverty line in industrialized countries, and which we define for reasons we give below as an individual 'development threshold'. Indeed, it is precisely because there is already this substantial population of developing country citizens—who, although (mostly) not wealthy, are not poor—that we consider it necessary to look within countries to properly estimate the 'capacity' to pay for climate policy (mitigation and adaptation).

In the remainder of this section we will describe in some detail the method that we use to calculate capacity and responsibility, and the way we combine them into a single 'Responsibility and Capacity Indicator', or RCI. Behind all the technical detail, however, it is important to understand that the guiding principle is simple: those who have more capacity to pay, and those who have more responsibility for creating the problem, should contribute more to solve it. This, we maintain is the ethical consensus that underlies the UNFCCC's recognition of 'common but differentiated responsibilities and respective capabilities'. What we have tried to do is to quantify and combine them in a way that adequately protects the poorest and least responsible from bearing the costs of climate policy; this is ultimately what we mean by the 'right to development'.

Defining and combining capacity and responsibility

We offer a very simple definition of capacity: for an individual, capacity is the income a person has over a specified (PPP adjusted) 'development threshold'. We leap over a stunning amount of complexity here, but the point is that there is a commonly held view of the difference between 'necessities' and 'luxuries'.[10] Capacity in this context is simply the ability to reduce one's consumption without sacrificing necessities. Thus it is a necessarily moral concept, and one that is useful despite the difficulty of specifying a necessity/luxury dividing line with precision; therefore our threshold figure should be understood as a first approximation for purposes of illustration. Nonetheless, there is plenty of evidence supporting the distinction (such as the continued support for progressive taxation).

We have picked for our 'central case' a development threshold of $7,500 per capita, in PPP adjusted dollars. This is obviously much higher than the $1 or $2/day threshold that typically defines 'absolute' poverty; rather, it resembles a poverty threshold in an industrialized country.[11] Keep in mind that, to the extent that PPP adjustments achieve their goal, this figure represents the purchasing power of $7,500 in the US, which is not very much. In terms of energy use, food

---

[9] Note that we do not intend to imply by this that 'development' is *simply* a matter of economic growth and high per capita incomes, but that discussion would take us far afield.

[10] Developing-country negotiators reference this distinction in speaking of 'luxury emissions' versus 'subsistence emissions' at international climate talks (Embassy of the People's Republic of China in Australia 2007), following Agarwal and Narain (1991) who first coined the expression.

[11] Pritchett (2006) argues that a global poverty line—as opposed to an extreme poverty line—should be on the order of $6,000 PPP adjusted; our $7,500 figure is roughly 125 per cent of this level.

Table 1. Sensitivity of covered population and measured capacity to alternative definitions of 'development threshold' (see main text)

	Population over threshold		*Capacity*	
Development threshold ($PPP adjusted, 2005)	*Total (billions)*	*Percentage of global*	*Total ($ trillion, $PPP adjusted, 2006)*	*Percentage of gross world product (GWP)*
6,000	2.1	33%	40.4	62
7,500	1.8	28%	37.1	57
9,000	1.6	25%	34.3	53

consumption, medical care and so forth, it goes only slightly beyond bare necessities. Indeed, in the industrialized countries, persons earning this amount or less are not only typically exempted from income taxes, but are eligible for subsidies of various sorts from the national treasury.[12]

Our point, however, is not that this threshold or any other is 'correct'. In Table 1, we show for our $7,500 threshold, and for slightly lower and higher figures, what fraction of the world's population falls above or below the threshold, and what fraction of the world's capacity is held by those above or below it.[13] It shows that the capacity of the 1.8 billion persons (roughly 28 per cent of the world's population) with incomes over $7,500 amounts to about 57 per cent of gross world product (GWP). Lowering the development threshold to $6,000 would increase the number of people above the threshold from 1.8 to 2.1 billion (a 17 per cent increase) but only increase the measured capacity from 57 per cent to 62 per cent of total GWP (a nine per cent increase). If, for example, one were trying to raise a sum equal to one per cent of GWP (about $590 billion in 2006) by an equal percentage tax on those with capacity, this would mean—for a $7,500 threshold—a tax rate of about 1.6 per cent on all income over $7,500, whereas for a $6,000 threshold, one could have a tax rate of 1.45 per cent on all income over $6,000. With the lower

[12] It is important to keep in view that $7,500 is an individual, not a family, threshold. Thus, threshold income for a family of four in the US would be a more realistic $30,000. However, since the poverty line for a family of four is not the same as four times the poverty level for an individual, no individual threshold can precisely capture the 'capacity' of differently-sized households. Furthermore, national plans could (and should, in our view) be designed so that the *actually taxed* threshold income is considerably higher than the *global* threshold we use to calculate countries' aggregate obligations. For instance, people with over a million dollars' annual income could certainly assume some of the obligations of individuals with incomes between $7,500 and, say, $50,000.

[13] We have used countries' per capita income and their Gini coefficients, which are widely accepted indicators of income inequality, to calculate countries' above-threshold income. We assume that national income can be modelled as a lognormal distribution, which (compared to the bell-curve shape of the well-known 'normal' distribution), is 'pushed to the left' and has a long right-hand 'tail' (see Kemp-Benedict 2001). Using this model, one can estimate how many people in each country are in each 'income bracket', and from this calculate capacity. Inequality matters because in a country with (say) a per capita income of $5,000, if that income were distributed perfectly evenly, the country would (by our definition) have no capacity; as rising inequality shifts income towards those whose individual income is over the threshold, the country's measured capacity rises.

threshold someone earning $7,500 per year would now pay around $20 instead of nothing, and someone earning $50,000 would pay about $640 instead of $670. Of course, whether this is desirable is a value judgement, but what is ultimately at issue is simply that if the income threshold is lowered, poor people must pay more for mitigation and adaptation so that rich people can pay less.

The way in which we calculate obligation on the basis of the distribution of income and emissions *within* countries means that setting the threshold in one place or another doesn't affect whether countries *as a whole* are above or below it; a country will always have obligation proportional to the capacity and responsibility of the fraction of its population, however large or small, that is above the line. This, we suggest, is a key political requirement of a viable and politically acceptable global climate regime. At present, Kyoto's complete exemption of non-Annex I countries from all cost-bearing obligations means that economic elites in poor countries can hide behind their own poor (Baer et al 2007, 66).

The second component that enters into the calculation of obligations in the GDRs framework is 'responsibility', defined broadly as 'contribution to the problem of climate change'. Allocation of burdens on this basis is thus an implementation of the widely endorsed 'polluter pays' principle. There are several ways to convert this general definition into a specific formula; one could count contributions to total GHG emissions, contribution to the remaining stock of GHGs in the atmosphere, or contribution to current or future temperature change, and one could also include or exclude various global sinks, such as forests. One could also choose different gases, different sources or different time periods. Thus any single definition has an element of arbitrariness, and countries could be expected to advocate definitions that reduce their relative responsibility. We choose *per capita emissions of* CO_2 *from fossil fuel consumption since 1990,*[14] a definition that gives greater responsibility to the developing countries relative to longer time periods, but gives greater responsibility to the developed countries than they would have if emissions from deforestation had been included. As a matter of necessary simplification, we assume that historical emissions are distributed in exact proportion to income; thus a fraction of emissions corresponding to 'below the threshold' income is excluded from our accounting.

Finally, to calculate the 'Responsibility and Capacity Indicator' (RCI) that we ultimately use to define a country's share of global obligations, we combine the measures of responsibility and capacity using the formula $RCI = R^a \bullet C^b$ with the restriction that $a + b = 1$. In our base case we set $a = 0.4$ and $b = 0.6$, which weights capacity somewhat more than responsibility. These could be weighted differently, but because some high emitters are poor we have assigned a greater weight to capacity here in the belief that it is fairer for rich countries with low emissions to pay more. Again, C (capacity) is 'income over the threshold,' and R (responsibility) is cumulative per capita emissions from fossil fuels, starting in 1990.

---

[14] Because the First Assessment Report of the IPCC was published in 1990, it is commonly argued that policy-makers have been informed of the climate change problem—and thus nations have moral responsibility—dating from that year.

Results

Table 2 shows, for selected countries and groups of countries, the results of our calculations. We include all of the 'plus 5' developing countries but not all of the G8.[15] We also include the Least Developed Countries (LDCs) and the Annex I and non-Annex I countries.

For each country or group we show, in percentage terms, its share of global population, global income (PPP adjusted), 'capacity' as we defined it above, cumulative fossil fuel emissions from 1990, 'responsibility' as we defined it above, and the combined RCI. Data is projected to 2010 using growth estimates from the International Energy Agency (2007) to reflect more accurately the distribution of income and emissions relevant for post-2012 burden sharing. Table 2 shows that, for all of the limits to the Annex I/non-Annex I distinction, Annex I still includes most (around 75 per cent) of the aggregate capacity and responsibility, despite comprising a small fraction (19 per cent) of the world's population. Together the US and the EU account for more than 55 per cent of calculated global obligations, as measured by the RCI, with the US alone accounting for 32 per cent. By contrast all non-Annex I countries, with 81 per cent of global population, account for only 25 per cent of global obligations. This is not a trivial share, however, as asking for it to be waived would significantly raise everyone else's 'bill'; with literally hundreds of billions of dollars a year at stake, 25 per cent is, as they say, 'real money' and represents a significant obligation for the wealthy in developing countries.

It is useful to go a little further and to consider the application of the GDRs 'obligation formula' to stylized models of fund-based and cap-and-allocate[16]-based implementation. Firstly, as we suggested above, we consider the possibility that the 'costs' of climate change policy (both mitigation and adaptation) could easily be one per cent of GWP—roughly $700 billion in PPP terms in our 2010 projection—which could hypothetically be paid by allocating that amount as a 'tax' on countries in proportion to their RCI. We show in Table 3, in addition to the total 'bill' a country would receive under this allocation, the percentage of people in each country who are over the 'threshold' per capita income. Then, using this fraction, we show the 'per taxpayer' bill on the assumption that only those over the threshold actually pay the tax.

Of course the GDRs framework could not mandate how a country's bill was to be allocated and collected internally, but certainly the political intent is to ensure that the international system lends legitimacy to the demands of poor people to fair treatment in their national policies.

Table 3 presents the results of a highly stylized model of a global assignment of obligations, yet it is intended to be the basis for evaluating concrete proposals. We offer here a defensibly fair allocation of a dollar-denominated global funding requirement, in national and per capita terms, and those who find our arguments

---

[15] The G8+5 group was formed in 2005 when Tony Blair invited the heads of state of five of the largest developing countries (Brazil, China, India, Mexico and South Africa) to the G8 meeting in Gleneagles, Scotland, in part to address climate change.

[16] We use the term 'cap and allocate' rather than 'cap and trade' to emphasize that, although (theoretically) economic efficiency is achieved by trading, the allocation of permits is what determines the fairness of the system and is a primary, not secondary, consideration.

Table 2. Key data and indicators for selected countries and groups of countries, based on 2006 data projected to 2010 (see main text; note that all numbers are in percentage terms)

	Per cent share (projected to 2010)					
	Global population	*Global income*	*Global capacity*	*Cumulative emissions since 1990*	*Global responsibility*	*Global RCI*
US	4.6	20.7	29.7	23.3	33.9	31.8
EU (27)	7.2	21.6	27.9	15.9	20.5	24.8
Russia	2.0	3.2	2.9	6.3	5.9	3.9
Brazil	2.9	2.8	2.3	1.4	1.2	1.8
China	19.7	12.5	5.9	15.7	7.5	6.6
India	17.2	5.2	0.8	4.2	0.7	0.8
Mexico	1.6	2.0	1.8	1.5	1.4	1.7
South Africa	0.7	0.7	0.6	1.6	1.4	0.9
LDCs	12.5	1.5	0.1	0.6	0.0	0.1
Annex 1	18.8	57.2	75.1	56.5	73.4	74.6
Non-Annex 1	81.2	42.8	24.9	43.5	26.7	25.4

Table 3. Characteristics of a fund-based system in which one per cent of GWP is raised by an equal percentage tax on the income of all persons over the $7,500 'development threshold' (see main text)

	Total income (billion $ PPP projected 2010)	*Total capacity (billion $ PPP projected 2010)*	*Percentage of global RCI*	*Bill at 1% of GWP ($ billion PPP)*	*Per 'taxpayer' bill at 1% of GWP ($PPP)*
US	14,226	11,909	31.8	219	733
EU (27)	14,845	11,192	24.8	171	412
Russia	2,173	1,166	3.9	27	228
Brazil	1,891	926	1.8	12	166
China	8,567	2,345	6.6	45	124
India	3,541	321	0.8	5.2	65
Mexico	1,368	730	1.7	11.5	199
South Africa	502	257	0.9	6.2	320
LDCs	1,018	36	0.07	0.4	57
Annex 1	39,357	30,112	74.6	513	447
Non Annex 1	29,400	9,973	25.4	175	169
World	68,757	40,085	100.0	688	315

plausible can compare these numbers[17] to those derived from other proposals to address the inevitable needs for some international funding of mitigation and adaptation. Of course, if total obligations turned out to be 0.1 per cent or five per cent of GWP, our formulas could still be applied; we picked the one per cent example in part because it's easy to multiply, not because it is necessarily our best estimate of the total 'bill' to be paid.

[17] Note that by 'these numbers' we don't mean narrowly the results in this particular table, but those that are derived from plausible variations in key parameters like the development threshold and the weighting of responsibility and capacity.

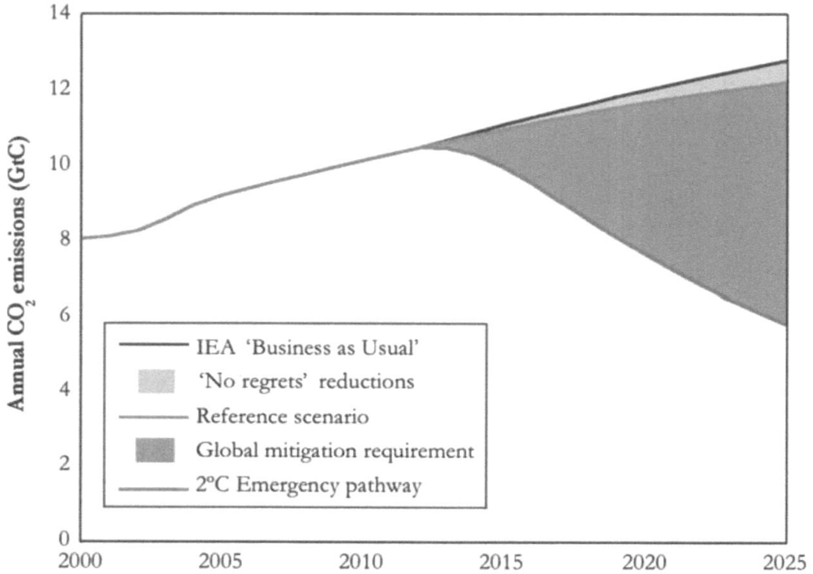

Figure 2. Global CO_2 emissions trajectory showing 'no regrets' reductions (small wedge) and the global 'mitigation gap' (large wedge) associated with a global 'emergency pathway'

Next we consider the implications of using the RCI as the basis of a cap-and-allocate system, in which hypothetically each country is allocated tradable emissions permits consistent with a precautionary global emissions trajectory. Figure 2 shows the global mitigation requirement or 'mitigation gap' between a hypothetical baseline and the precautionary 'emergency pathway' shown in Figure 1. We show two different baselines, an upper 'Business as Usual' (BAU) baseline and a lower 'reference scenario' baseline which takes into account negative and zero-cost 'no regrets' mitigation options. The larger wedge represents the emissions reductions that must be paid for by someone; the 'no regrets' wedge represents reductions that are presumed profitable or of no net cost to the country that makes them, and thus don't represent costs to be allocated.[18]

Figure 3 shows the 'mitigation gap' from Figure 2, divided into wedges proportional to the RCI for the indicated country or group of countries. This is analogous to the 'technology wedges' defined by Pacala and Socolow (2004), except that it shows the share of reductions to be accounted for by a country, rather than a technology. In this regard, it resembles the way in which the so-called 'Brazilian proposal' would have allocated global reductions, by dividing a mitigation objective in proportion to a single indicator of national 'shares'.[19]

---

[18] The 'Business as Usual' baseline is derived from projections of the International Energy Agency (2007), modified for emissions reductions through 2012. 'No regrets' reductions are based on the estimates of McKinsey and Company (Enkvist et al 2007) and the IPCC (2007b, Table TS 15).

[19] The Brazilian Proposal of 1997 allocated shares of emissions reductions in proportion to countries' contributions to global temperature change, and took account of capacity by excluding non-Annex I countries from such shares. Otherwise it is perhaps the closest relative of the GDRs approach (Meira Filho et al undated).

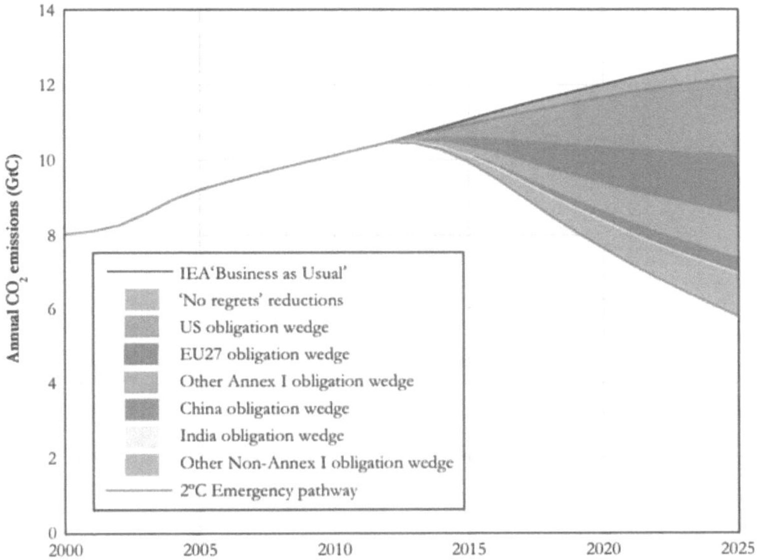

Figure 3. National 'reduction obligation wedges' carved out of global 'mitigation gap' in proportion to national (or regional) RCI. *Note*: the legend is in the same order as the wedges, and the 'India' wedge is effectively a line

In order to translate the 'obligation wedges' shown in Figure 3 into an actual allocation of emissions permits, one needs additionally to estimate the national baseline emissions trajectory; a country's obligation is then calculated by subtracting its percentage share of the global mitigation requirement from its baseline. We show such a calculation for the US in Figure 4, and for China in Figure 5. Again, the BAU baseline and 'no regrets' reductions should be taken as merely indicative; the key assumption is simply that a country's baseline

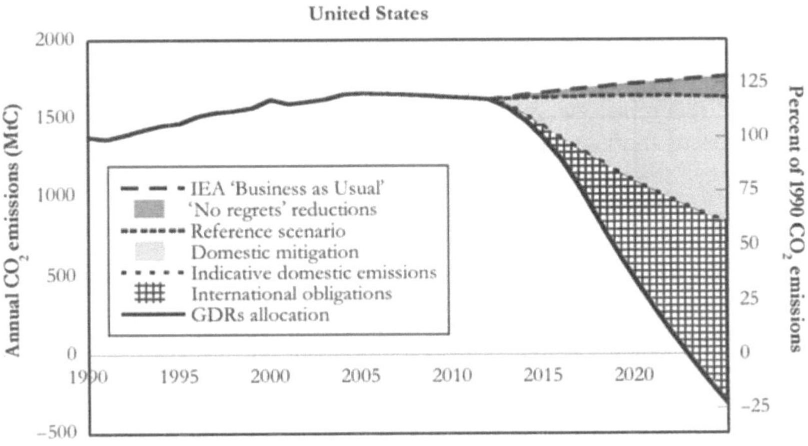

Figure 4. Hypothetical distribution of emissions reductions and obligations for the US under the GDRs framework (see main text)

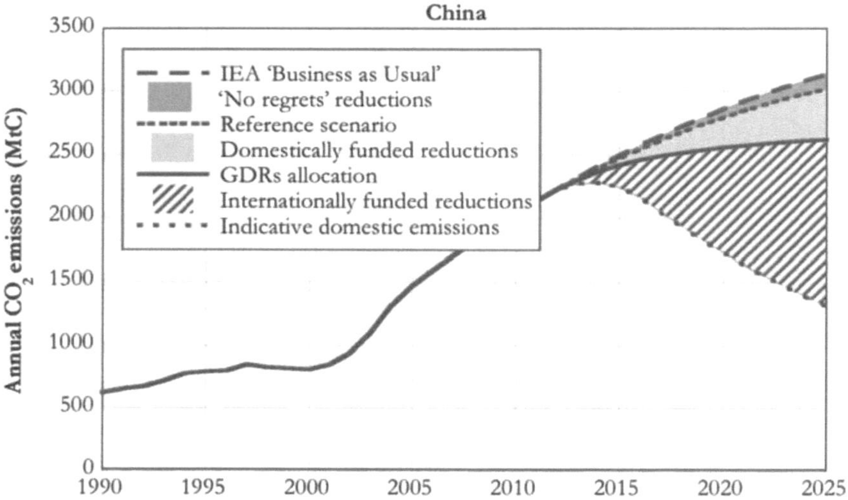

Figure 5. Hypothetical distribution of emissions reductions and obligations for China under the GDRs framework (see main text)

allocation, prior to its mitigation obligation, would be less than historical 'BAU' growth due to the availability of 'no regrets' mitigation options.

What we see for the US is a hypothetical 'no regrets' wedge that expands gradually between 2012 and 2025. Beneath that are two additional wedges: a wedge of *domestic mitigation*, calculated in this case to decline at six per cent annually, reaching 90 per cent below 1990 levels in 2050 (consistent with the Annex I target in Figure 1, as well as with the proposal of Al Gore and others for US reduction targets) and a cross-hatched wedge of *international reduction obligations*. Together these two add exactly to the US share of reductions shown in Figure 3, but they are shown against the national baseline rather than the global baseline. What is crucial is that by 2025, US total obligations are substantially larger than very optimistic domestic reductions. The obligations are larger than the US' total current emissions; in effect, a *negative permit allocation*. That is, in order to comply with its obligations, the US must not only purchase permits for its actual emissions, but also enough to cover the 'gap' between its negative allocation and its emissions. Although this may seem counterintuitive, it is not in principle different than being given an allocation that is smaller than a country's actual emissions; it simply means that more permits must be bought to achieve compliance.

The complementary situation for China, whose obligations are substantially *less* than the (assumed) globally cost-effective level of reductions, is shown in Figure 5. We assume here that in a world with an efficient trading regime (remember that this is only a model!), physical reductions in each country would still approximate the necessary global rate. Thus we show overall emissions (the lower line) declining at six per cent annually; the middle wedge represents the emissions reductions that China must pay for (equivalent to China's wedge in Figure 3), and the (substantially larger) striped wedge represents emissions reductions in China paid for by other countries.

Such payment obligations are precisely why one might expect that a framework like GDRs will not easily find favour in the US or other wealthy, high-emitting countries, most of which will also have negative allocations before 2025. However, if one also accepts that it is *necessary* that physical emissions drop as rapidly in China as we show, the fact remains that *someone* will have to pay for them. What we have done is create a method by which all the *people* in the world are effectively allocated obligations proportional to their capacity and responsibility; the international transfers that follow are simply a consequence of the actual distribution of wealth and emissions between and within countries.

We will comment on the political issues raised by the framework in the conclusion of the article. Before that, however, it is worth taking one additional detour to compare the GDRs framework with the most prominent 'equity framework' to date, which is allocation of emissions rights on an equal per capita basis—in particular, the 'Contraction and Convergence' (C and C) framework promoted by the Global Commons Institute in the UK (Meyer 2000). In C and C, emissions allocations 'converge' from allocation in proportion to historical emissions (grandfathering) at the beginning to allocation in proportion to population (equal per capita) at the end of a designated period.

Advocates of C and C argue that the early allocation of 'surplus' permits to low-emitting countries would allow them to fund low-carbon development. However, under global trajectories stringent enough to have a high likelihood of staying below 2°C, like our 'emergency pathway', many developing countries will have to begin reducing emissions immediately or within the next decade, long before they reach the per capita income or emissions levels of the industrialized countries. Furthermore, the system makes no distinction as to capacity between richer and poorer high-emitting countries (for example Belgium and Kazakhstan, both at 3.6tC per capita in 2005) and richer and poorer low-emitting countries (for example Argentina and North Korea, both around 1tC per capita in 2005).[20] In contrast, the GDRs framework moves beyond per capita frameworks to fairly incorporate capacity as well as responsibility into the calculations of obligations, in order to actually protect the 'right to development'.[21]

Implementation: how GDRs might be used in a post-Kyoto context

By specifying global mitigation obligations in terms of precautionary ecological criteria and offering a formula for quantifying the obligations of states according to their historic responsibility and capacity, the GDRs framework might be used to rectify the shortcomings of the Kyoto Protocol. Whereas the UNFCCC's stated objective is to limit anthropogenic GHG emissions so as to prevent climate destabilization in a manner that takes account of different state capacities, the Protocol's modest reductions and emphasis on flexibility mechanisms, and its failure to offer a principle sufficiently specific to determine obligations, undercut

---

[20] Emissions data, like most of that used in this analysis, are from the US Energy Information Administration, <http://www.eia.doe.gov/environment.html>.

[21] For a more detailed review of Contraction and Convergence and several other equity-based climate policy frameworks, see Baer and Athanasiou (2007).

its primary purpose. Thus we next consider ways in which GDRs might serve as the basis for a post-Kyoto agreement.[22]

We assume that recent findings of climate scientists, as well as the findings of the 2006 Stern Review (which estimated that the costs of *in*action on climate will dramatically exceed the costs of action) have enhanced the interests of most states in adopting a rigorous post-Kyoto instrument in which all countries, including developing ones, assume more substantive mitigation obligations. Moreover, enhanced perceptions of vulnerability in the global north are also likely, as Biermann (2002) suggests, to increase the power of southern countries in bargaining for equitable treatment in the form of financial assistance, as a condition for their assumption of mitigation obligations. Roberts and Parks argue that dealing southern nations 'a fair and predictable hand' will almost certainly lead to 'an immediate and lasting payoff in terms of international environmental cooperation' (2007, 229–231).

How then could GDRs be brought into the negotiations process to support enhanced cooperation? The key point is that the central quantitative 'result' of the GDRs framework is simply *national shares based on the aggregated capacity and responsibility of individuals*. In this regard it can be used in almost any context within the negotiations in which obligations need to be divided between countries. Mitigation and adaptation are plainly the two 'supercategories', but other types of quantifiable obligations could be divided similarly.

Concerning mitigation, it is useful to consider the two 'ideal type' models we described in the previous section as a starting point for implementation of the GDRs framework. As we suggested, our formula could be used either to define national shares of a global mitigation obligation measured in tons, in which case the key step is to define the allocation of emissions rights in terms of reductions below a particular baseline, or national shares of a global mitigation obligation defined in monetary terms. The former is simple to envision; the system would be essentially the same as the existing Kyoto Protocol's allocation of 'assigned amounts', which fairly closely follow the textbook example of a cap-and-trade system. The difficulties of moving from an industrialized-countries-only system to a global system would be many, and we discuss them in further detail below. But those issues are in general the same as those that will be associated with any extension of Kyoto to most or all of the nations of the world, regardless of the basis of the allocation.

The second case of a central global mitigation fund is also easy to imagine in theory, but harder to imagine soon coming into being. The World Bank has created a 'Prototype Carbon Fund' and a 'Community Development Carbon Fund' which can support various kinds of mitigation activities, but they are voluntarily funded and small relative to the investments implied by a global 'emergency pathway'. Most analysts think that an effective international taxing authority with a mandate of several hundred billion dollars annually is not likely to win the necessary support of sovereign nations.

---

[22] We use 'post-Kyoto' as shorthand for 'subsequent to the first commitment period of the Kyoto Protocol', which runs from 2008 to 2012. The ongoing negotiations under the so-called 'Bali Action Plan' imply that the Protocol will likely remain in effect, but with new targets established for a second commitment period.

Although the GDRs framework offers the prospect of a global Kyoto-style allocation of tradable permits, and of transcending the artificial Annex I and non-Annex I categories, there are many difficulties along that path. First, the implementation of national mitigation policies capable of supporting the appropriate distribution and trading of a national permit allocation is likely beyond the near-term capacity of many developing countries. Similarly, questions have been raised about the legitimacy of some governments as agents to sell emissions permits on behalf of their citizens (for example Singer 2002). And of course it is problematic to assume that a market for an artificial commodity like emissions permits will smoothly mediate between the demand for and supply of emissions reductions; the volatility of the European Emissions Trading System in its early years should provide ample caution in this regard.

The question remains then what kinds of 'mix and match' institutions might actually meet the mandate implied by the GDRs allocation of obligations. Here it is helpful to consider the current status of the negotiations under the UNFCCC and Kyoto Protocol, as they present some concrete opportunities and constraints with regard to possible institutional requirements and solutions.

Critically, the 'Bali Action Plan' implies a continuation of the broad division between Annex I (industrialized) and non-Annex I (developing) countries. The need for increased mitigation in developing countries was acknowledged in a controversial phrase calling for 'Nationally appropriate mitigation actions by developing country Parties in the context of sustainable development, supported and enabled by technology, financing and capacity-building, in a measurable, reportable and verifiable manner' (UNFCCC 2008b, paragraph 1(b)(ii)). Notably, the phrase 'measurable, reportable and verifiable' ambiguously applies to both sides of the agreement, the 'nationally appropriate actions' in non-Annex I countries, and the 'technology, financing and capacity-building' to be provided by Annex I countries. The GDRs framework offers some insight into both the scale (though not the details) of 'nationally appropriate' mitigation activities in developing countries, and how any funding for substantial monetary or monetarily valued transfers might be fairly allocated.

Beyond that, the questions of governance and procedures for choosing mitigation options are general ones for which the principles of the GDRs framework offer little specific insight. Indeed, one criticism of the GDRs framework is that it does *not* in fact offer solutions to critical institutional problems. It is ultimately premised on the assumption that massive northern capital flows will be necessary to fund mitigation activities in the South, but neither the experience of the Clean Development Mechanism (CDM) to date, nor the broader history of northern development assistance give grounds for optimism about the effectiveness of such flows were they to materialize. As we suggested at the beginning, however, such transfers are apparently a necessary condition of environmental adequacy; what GDRs offers is a possibility to take these transfers out of a simplistic north/south dichotomy by identifying, fairly and quantitatively, the obligations associated with the actual wealth and capacity of the rich and middle classes in poor countries. Beyond that, these problems are our problems but they are everyone's problems as well.

A point worth stressing here is that the GDRs framework, in a sense, is ahead of its time: it is intended as an architecture for a truly 'post-Kyoto' world, in which the simplistic divisions between Annexes have already been transcended and a global emissions budget adopted. However that time has plainly not yet come. Given the failure of the industrialized countries to meet Kyoto in either letter or spirit, the developing countries are insisting on maintaining the fundamental distinction regarding who gets quantified emissions targets and who does not. Our RCI can be easily used to show how the current Annex I/non-Annex I distinction is inadequate, but its solution of giving *all* countries *some* quantified obligations is not presently on the table. Furthermore, neither the US nor leading developing countries seem interested in proposing or accepting a global mitigation budget consistent with (for example) the 2°C threshold endorsed by the EU and others.

One possible response to this impasse would be the adoption of an abbreviated second commitment period, perhaps from 2013 to 2015. Maintaining the Kyoto distinctions for this short additional period would not greatly increase the burden on the wealthier countries, which are primarily exposed to Kyoto's mitigation costs. It would provide an opportunity for trust-building, since it is enough time to allow Annex I countries to show commitment to both domestic mitigation and international mitigation and adaptation obligations while non-Annex I countries proceed to integrate mitigation and adaptation into their development plans. Meanwhile (concurrently), negotiators could address the larger problems of global targets and comprehensive burden sharing without quite the urgency of the current 2009 deadline.

In the interim phase, GDRs, or something similar, could be used to allocate any obligations or responsibilities that can be quantified and aggregated. For example, the requirement to provide funding for adaptation has yet to be addressed in any principled manner; so far only voluntary commitments and levies on CDM projects in developing countries have been adopted. Other proposals (for example, an aviation levy) seem to rely on the appearance of *not* being tied to acknowledgement of legal liability on any nation's part. In contrast, GDRs offers one way to combine responsibility and capacity that is defensibly related to the obligation to pay for adaptation sufficient to protect likely victims from harm.

Finally, it is worth pointing out that the GDRs framework has implications for the allocation of costs within countries as well as between countries. The allocation of national shares is based on the aggregation of the responsibility and capacity attributed to (modelled) individuals within a country. There is an implicit tax bill (or permit allocation) for every hypothetical individual within a country, with an exemption to all income under the development threshold. If a country were to adopt GDRs as an international position, it would follow that the domestic allocation of costs should be at least equally progressive. This is not to say that the framework as embodied in international law would be able to mandate how nations should raise or distribute funds associated with any domestic policies. However, the privilege of participating in (say) an international trading regime could be conditioned upon some minimal conditions of fairness in the implementation of domestic policies and measures. Conditionalities of various kinds routinely act as 'gateways' for states' participation in international regimes; for example, participation and disclosure requirements must be met if CDM project credits are to be internationally tradable.

Conclusion

We have some confidence that a proposed international agreement based on the GDRs framework would (eventually) elicit the full participation of developing countries, including those, like India and China, with high aggregate but low per capita emissions, because it specifically protects their right to development. The commitment of these countries to mitigation obligations (with international financial assistance in many cases) is necessary not only for 'climate adequacy', but also to answer the objection of the US, in particular, to the exemption of developing countries from mitigation obligations under the Kyoto Protocol. The GDRs framework accommodates global carbon trading; the resulting global price for carbon would help solve the problem of 'runaway' investment and disadvantages to high-obligation countries in trade, since emission permits at the global price would be required anywhere production is carried out. Carbon trading is favoured by most economists as the most cost-effective way to achieve emissions reductions (see for example Olmstead and Stavins 2006).

However, mitigation commitments on the part of China, India and similar developing countries would be insufficient, we believe, to elicit the near-term support of northern high emitters—especially the US—for a GDRs-based international agreement with a stringent (environmentally adequate) reduction scenario, despite near-universal recognition of the gravity of the climate crisis. This is quite simply because of the financial demands it would place on northern high emitters, whose often stated concern has been that such obligations would diminish economic growth. When the costs of mitigation and adaptation are placed in an appropriate context—such as in comparison to defence expenditures, or to the estimated costs of inaction, as in the Stern Review—they are neither unimaginable nor unmanageable (Stern 2006). But it is not only abstract 'economic growth' that is challenged by what we have proposed. High-emitting developed countries *could* finance mitigation and adaptation by reducing their defence spending proportionately, but at the perceived risk of threatening their positions in the world system—a particular problem for the US. They *could* adopt a suite of policies that provides and encourages the use of public goods and services rather than private consumption, and discourages some forms of private consumption altogether—in other words, a 'strong' version of 'Ecological Modernization'[23]—but this would challenge the freedom of corporations that has been structured into the institutions and rules of so-called 'neoliberal' global capitalism in recent decades.

As a cautionary example, den Elzen et al (2006) approach the question of whether parties *would* undertake the costs of adequate mitigation by constructing what they call '[t]he political willingness scenario for 2020'. The exercise sought to determine implications for long-term stabilization of emissions, given the reductions hypothesized to be politically acceptable between 2012 and 2020.

[23] Eckersley, among others, distinguishes among the substantial range of policy approaches that reside under the umbrella term 'Ecological Modernization'. The weak version 'amounts to little more than a cost minimization strategy for industry'. Alternatively, 'strong' ecological modernization is the product of a discursive, reflexive and critical *process* that translates equitable and precautionary values into deep structural changes in the economy, society and state; it is thus much more than an industrial policy (Eckersley 2004, 70–77).

The authors conclude that even a (long-term) stabilization of 400–450 ppm CO_2-e will be beyond reach, given their assessment of political willingness of the developed (Annex I) countries to bear the expense of mitigation up to 2020.[24] A GDRs-style adequate and equitable agreement certainly exceeds these bounds of 'political willingness'.

Nonetheless, we do not present GDRs as a mere thought exercise. We see considerable potential not only in global social justice NGOs, but also in the political movement of citizens nationally and globally, to secure a more just and sustainable future by demanding accountability from states, corporations and undemocratic global institutions. Spokespersons for the global justice movement have specifically called for a regulated and redistributive global capitalism to replace the current 'neoliberal' version that has maximized mobility and minimized the accountability and financial obligations of corporations to society (see George 2004). 'Taming' capitalism in this manner is a prerequisite for the implementation of a rigorous version of Ecological Modernization. Moreover, recent polls conducted by GlobeScan and the Program on International Policy Attitudes at the University of Maryland suggest that publics in both the US and China are receptive to climate adequacy and equity in ways that state managers are not: large majorities of *both* samples (the smallest majority was 59 per cent) agreed that climate change is a serious problem demanding major steps; that behavioural and lifestyle changes are necessary and that fuel prices must go up to encourage such changes; and that 'wealthy countries [should] agree to provide less-wealthy countries with financial assistance and technology' while both wealthy and poor countries agree to limit their emissions (Kull and Miller 2008, 14).

Sustainability and equity may indeed be beyond the capacity of the present neoliberal and Westphalian world order. However, the potential for nonlinear change is characteristic of any complex system, including the global political economy. We conceive GDRs, therefore, as part and parcel of a global social project whose realization is not yet in view but which, at least some evidence suggests, is under construction.

References

Agarwal, Anil A and Sunita Narain (1991) *Global warming in an unequal world: a case of environmental colonialism* (New Delhi: Centre for Science and Environment)

Baer, Paul and Tom Athanasiou (2007) 'A brief, adequacy and equity-based evaluation of some prominent climate policy frameworks and proposals', Global Issue Paper 30, Heinrich Böll Foundation, Berlin, <http://www.boell.de/alt/downloads/global/global_issue_paper30.pdf>, accessed 30 July 2008

Baer, Paul, Tom Athanasiou and Sivan Kartha (2007) *The right to development in a climate constrained world: the Greenhouse Development Rights framework* (Berlin: Heinrich Böll Foundation), <http://www.ecoequity.org/docs/TheGDRsFramework.pdf>, accessed 30 July 2008

Baer, Paul and Michael Mastrandrea (2006) *High stakes: designing emissions pathways to reduce the risk of dangerous climate change* (London: Institute for Public Policy Research),

[24] Den Elzen et al (2006) assume that the US reengages with the UNFCCC/Kyoto process, and reduces its emissions 15 per cent below 1990 levels by 2020.

<http://www.ippr.org/publicationsandreports/publication.asp?id=501>, accessed 30 July 2008

Biermann, Frank (2002) 'Power and interdependence in global environmental policy: explaining the new bargaining power of developing countries', Paper presented at 43rd International Studies Association Annual Conference, New Orleans, 24–27 March

Canadell, Josep G, Corinne Le Quere, Michael R Raupach, Christopher B Field, Erik T Buitenhuis, Philippe Ciais, Thomas J Conway, Nathan P Gillett, RA Houghton and Gregg Marland (2007) 'Contributions to accelerating atmospheric CO_2 growth from economic activity, carbon intensity, and efficiency of natural sinks', *Proceedings of the National Academy of Sciences*, 104:47, 18866–18870

Den Elzen, Michel JG, Niklas Hohne, Bernd Brouns, Harald Winlker and Hermann E Ott (2006) 'Differentiation of countries' future commitments in a post-2012 climate regime: an assessment of the "South-North Dialogue" proposal', *Environmental Science and Policy*, 10:3, 185–203

Eckersley, Robyn (2004) *The green state: rethinking democracy and sovereignty* (Cambridge, Massachusetts: MIT Press)

Embassy of the People's Republic of China in Australia (2007) 'China urges accommodation to "emissions of subsistence" by poor countries', 2 August, <http://au.china-embassy.org/eng/xw/t347202.htm>, accessed 31 July 2008

Enkvist, Per-Anders, Tomas Nauclér and Jerker Rosander (2007) 'A cost-curve for Greenhouse gas reduction', *McKinsey Quarterly*, 2007:1, 35–45

George, Susan (2004) *Another world is possible if...* (New York: Verso)

Hansen, James (2007) 'Scientific reticence and sea level rise', *Environmental Research Letters*, 2:024002

International Energy Agency (2007) *World energy outlook 2007* (Paris: OECD/IEA)

IPCC (2007a) *Climate change 2007: synthesis report*, <http://www.ipcc.ch/ipccreports/ar4-syr.htm>, accessed 31 July 2008

IPCC (2007b) *Contribution of Working Group III to the Fourth Assessment Report of the Intergovernmental Panel on Climate Change*, <http://www.ipcc.ch/ipccreports/ar4-wg3.htm>, accessed 31 July 2008

Kemp-Benedict, Eric (2001) 'Income distribution and poverty: methods for using available data in global analysis', PoleStar Technical Note 4, <http://gdrs.sourceforge.net/docs/PoleStar_TechNote_4.pdf>, accessed 1 September 2007

Kull, Stephen and Doug Miller (2008) 'Let people lead on climate change', *The Progressive Populist*, 14:11, 14

Meira Filho, Luiz G, Gonzales Miguez and Jose Domingos (undated) 'Brazilian proposal for Kyoto', transl Annesandra de Avila Ribeiro, <http://ecen.com/content/eee3/Kyotoe.htm>, accessed 31 July 2008

Meyer, Aubrey (2000) *Contraction and convergence: the global solution to climate change* (Totnes, United Kingdom: Green Books Ltd)

Olmstead, Sheila M and Robert N Stavins (2006) 'An international policy architecture for the post-Kyoto era', *American Economic Review*, 96:2, 35–38

Pacala, Stephen and Robert Socolow (2004) 'Stabilization wedges: solving the climate problem for the next 50 years with current technology', *Science*, 305:5686, 968–972

Page, Edward A (2006) *Climate change, justice and future generations* (Northampton, Massachusetts: Edward Elgar)

Pritchett, Lant (2006) 'Who is not poor? Dreaming of a world truly free of poverty', *World Bank Research Observer*, 21:1, 1–23

Roberts, J Timmons and Bradley C Parks (2007) *A climate of injustice: global inequality, north-south politics, and climate policy* (Cambridge, Massachusetts: MIT Press)

Science Daily (2007) 'Arctic Sea ice minimum shatters all time record low, report scientists', 21 September, <http://www.sciencedaily.com/releases/2007/09/070920160226.htm>, accessed 12 June 2008

Singer, Peter (2002) *One world: the ethics of globalization* (New Haven, Connecticut: Yale University Press)

Spratt, David and Philip Sutton (2008) 'Climate "code red": the case for a sustainability emergency', <http://www.climatecodered.net>, accessed 31 July 2008

Stern, Nicholas (2006) *Stern review on the economics of climate change* (London: Government Economic Service), <http://www.sternreview.org.uk>, accessed 31 July 2008

UNFCCC (1992) *United Nations framework convention on climate change*, FCCC/Informal/84, <http://unfccc.int/resource/docs/convkp/conveng.pdf>, accessed 31 July 2008

UNFCCC (2008a) 'Report of the Conference of the Parties on its thirteenth session, held in Bali from 3 to 15 December 2007. Part 1: Proceedings', UNFCCC/CP/2007/6, 14 March, <http://maindb.unfccc.int/library/view_pdf.pl?url=http://unfccc.int/resource/docs/2007/cop13/eng/06.pdf>, accessed 31 July 2008

UNFCCC (2008b) 'Report of the Conference of the Parties on its thirteenth session, held in Bali from 3 to 15 December 2007. Addendum. Part 2: Action taken by the Conference of the Parties at its thirteenth session', UNFCCC/CP/2007/6/Add.1, 14 March 2008, <http://maindb.unfccc.int/library/view_pdf.pl?url=http://unfccc.int/resource/docs/2007/cop13/eng/06a1.pdf>, accessed 31 July 2008

Conclusion: constructing the climate regime

Paul G Harris

What is it about climate change that has caused governments to react so slowly? One answer is found in *ideas* about climate change and their effect on politics. The notion that ideas matter in international affairs is not new, as reflected in the constructivist literature cited in preceding essays. In the case of climate change, problems lie in the multiple identities it evokes. For example, some officials view climate change as a minor issue, less important than other concerns. Alternatively, some people and governments recognize that in the medium and long-term climate change will have profoundly adverse consequences. For them, it is a priority. For other officials, climate change is about environmental stewardship or even morality, and many nongovernmental actors—alongside developing world governments—view it as a matter of fairness. In short, the spectrum of views on climate change is very wide. What is more, some people see climate change from many perspectives: it may be simultaneously perceived as a short-term economic concern for one's own state, a question of international distributive justice in the medium term, and a vital matter of national and global security in the long term.

Much as the climate is changing, the idea of climate change is a moving target that varies as the science shifts. With improved understanding of the causes and impacts of climate change, more governments recognize that inaction is no longer a credible option. Many people now identify with the problem and want to act accordingly. However, the *kinds of ideas* propagated internationally remain a central factor shaping outcomes. The climate regime has been constructed on a weak ideational foundation based on out-of-date (if quite recent) science sullied by 'climate sceptics' and popular media. Just as more governments are starting to call seriously for greenhouse gas cuts of 50 per cent (and sometimes more) below recent levels by mid-century, it becomes clear that cuts of 100 per cent *or more* (through carbon sequestration) will be required. Such cuts are not on the political cards, domestically or internationally.

Another problematic feature of climate politics is the many competing levels of political bargaining and action. It is a matter of local, domestic and international politics (and economics), just as it is affecting people locally, nationally and globally. The multiple conceptions of climate politics, and the identities that arise as a consequence, operate in overlapping layers of conflicting actors and processes—making agreement on how to deal with climate change dizzyingly complicated. Indeed, the politics of climate change make the politics of other notoriously complicated issues, such as international trade and arms proliferation, look relatively simple.

What will be required to construct a new, post-Kyoto climate regime amid this interplay of international and domestic politics? First and foremost, a much greater sense of urgency will have to sink in. Moreover, this sense of urgency has to be shared more widely—and very soon. Once enough people and governments are convinced that their homelands cannot avoid the adverse effects of climate change, and that acting now is less costly than acting much later, the pace of policy innovations will pick up. Additionally, international justice, and even cosmopolitan justice, will have to be seen as central to the climate regime. International justice is required to persuade major developing countries to place some limitations on their future emissions; cosmopolitan justice is required to ensure that growing legions of wealthy people around the world do not pollute at the expense of everyone, especially the poor, in the future.

If climate catastrophe and all it entails for the world is to be averted, we must make the transition with all possible haste to post-carbon societies. That such a statement seems manic to some people shows that the *idea* of climate change and the *reality* of it remain much too far apart. Even the consensus-based scientific findings of the Intergovernmental Panel on Climate Change, and especially independent scientific assessments of future effects of climate change, leave no doubt about the severity of the problem. To be sure, world leaders no longer stridently deny the reality of climate change. However, the ideas that are driving climate politics remain dated and lacking in immediacy. Half-measures and decades-long debates about which states should do what first have become, both practically and ethically, perverse.

Are the politics of climate change up to the task? That remains an open question. Scientists have passed judgement, and now it is up to leaders, officials and diplomats, as well as industry and citizens themselves, to find efficient, effective and equitable means of responding to climate change expeditiously. As the melting of the world's glaciers gathers pace, it is time for politics to catch up.

Index

Page numbers in *Italics* represent tables.
Page numbers in **Bold** represent figures.